AF879786

Nano and Micro Plastics

Nano and Micro Plastics

The Dangers We Face

Confronting the Invisible Crisis

Robert Sudar

Nano and Micro Plastics: The Dangers We Face, Confronting the Invisible Crisis Copyright © 2026 by Robert Sudar.

All Rights Reserved. No part of this publication may be reproduced, distributed, or transmitted in any form or by any means, including photocopying, recording, or other electronic or mechanical methods, without the prior written permission of the publisher, except in the case of brief quotations embodied in critical reviews and certain other noncommercial uses permitted by copyright law.

Published by RCS Experience Sheridan, WY https://rcs-experience.com/books

ISBNs Paperback: 978-1-971877-00-6

Library of Congress Control Number: 2026902766
U.S. Copyright Office Registration: TXu 2-506-703
Effective Date: July 15, 2025

Ordering Information: Quantity sales. Special discounts are available on quantity purchases by corporations, associations, and others. For details, contact the publisher at the address above.

Disclaimers: The information in this book is for educational and informational purposes only and is not intended as medical, legal, or professional advice. While every effort has been made to ensure the accuracy of the information herein, the author and publisher assume no responsibility for errors or omissions. Always seek the advice of a qualified professional regarding personal health or environmental safety.

Printed in the United States of America. First Edition: 2026

To the memory of Jimmy Boyd, whose tireless dedication to advancing men's health awareness inspired a movement that continues to this day.

And to the following individuals, whose work, insights, and dedication compelled me to investigate this crisis and inspired me to write this book:

Armin Brott
Bill Harrington
Warren Farrell
Tracie Snitker
Ana Fadich-Tomsic
Michael Leventhal

Your voices and contributions have illuminated the path toward better understanding and support for men's and boys' health. This book is a testament to your influence.

Introduction

Confronting the Invisible Crisis

For more than half a century, nano and microplastics have been silently accumulating across every corner of the planet. Initially overlooked, these tiny fragments—often less than a millimeter in size—are now embedded in our food, water, air, soil, and even our own bodies. From the depths of the Mariana Trench to the peaks of the Himalayas, microplastics are found in rainfall, infant formula, placental tissue, and the air we breathe at home. No environment is untouched, and no human is entirely unexposed.

This book is your complete A-to-Z guide to under-standing nano and microplastics: where they come from, how they affect human and planetary health, and what you can do about them. It offers scientific insight, urgent health alerts, and actionable steps to protect yourself and your loved ones in an increasingly toxic environment. More importantly, it illuminates the path to detox—not just personally, but globally.

Awareness is the first step. Around the world, countries have begun to respond: France has mandated microfiber filters on washing machines, Kenya has banned single-use plastic bags, Norway has upgraded wastewater treatment for microplastic capture, and over 100 nations have banned microbeads in cosmetics. These efforts are vital—but not enough.

We must now accelerate solutions that remove existing toxicity from the planet. This includes large-scale wastewater upgrades, plastic-eating enzymes, advanced filtration systems, and circular design strategies that stop pollution before it begins.

At the personal level, you'll learn how to detox your home, reduce exposure through diet and hygiene, and support your body's defenses through lifestyle and nutrition. Whether you're a policymaker, a parent, or a concerned citizen, this book empowers you to act.

Microplastics may be invisible—but their impact is not. The time to face them is now.

Table of Contents

Chapter 1: Understanding Nano and Micro Plastics

Introduction

In the twenty-first century, few environmental concerns have risen as swiftly in public awareness as nano and microplastics. Once an obscure issue relegated to niche scientific journals, these minute plastic particles have now entered the mainstream, appearing in headlines, documentaries, class-rooms, and policy debates. Their widespread presence in oceans, soil, food, and even human blood has galvanized researchers, activists, and policymakers into action[1]. But to address the issue meaningfully, one must begin with a precise and comprehensive understanding of what nano and microplastics are—how they're defined, how they're formed, and why they are both a scientific curiosity and a global crisis.

Definitions: Micro vs. Nano

Figure 1: MicroPlastic Particles on Fingertips
Source: European Union, 2025, CC BY 4.0 <https://creativecommons.org/licenses/by/4.0>, via Wikimedia Commons

Nano and microplastics are both tiny plastic particles, but they differ in scale, behavior, and potential biological effects. These terms are often used interchangeably in popular media, but in the scientific community, their distinctions are precise and important[2].

- **Microplastics** are typically defined as plastic particles measuring between 1 micrometer (µm) and 5 millimeters (mm). These can be further categorized into primary and secondary microplastics:
  - **Primary microplastics** are deliberately manufactured at microscopic size, such as exfoliating beads in cosmetics or industrial abrasives.[3]

- ◦ **Secondary microplastics** result from the breakdown of larger plastic objects through exposure to sunlight, wind, waves, or mechanical abrasion.[4]

- **Nanoplastics** are usually defined as plastic particles smaller than 1 micrometer, often down to sizes measured in nanometers (1 nanometer = one-billionth of a meter). These are more difficult to detect, study, or filter out, and they may behave more like colloidal particles or chemical contaminants than like solid debris.[5]

The exact cutoff between **nano** and **microplastic** isn't universally agreed upon, but the 1 µm threshold is commonly used[6]. Some researchers argue for a more fluid continuum based on function and mobility rather than fixed size boundaries.

Size Matters: Behavior and Risk

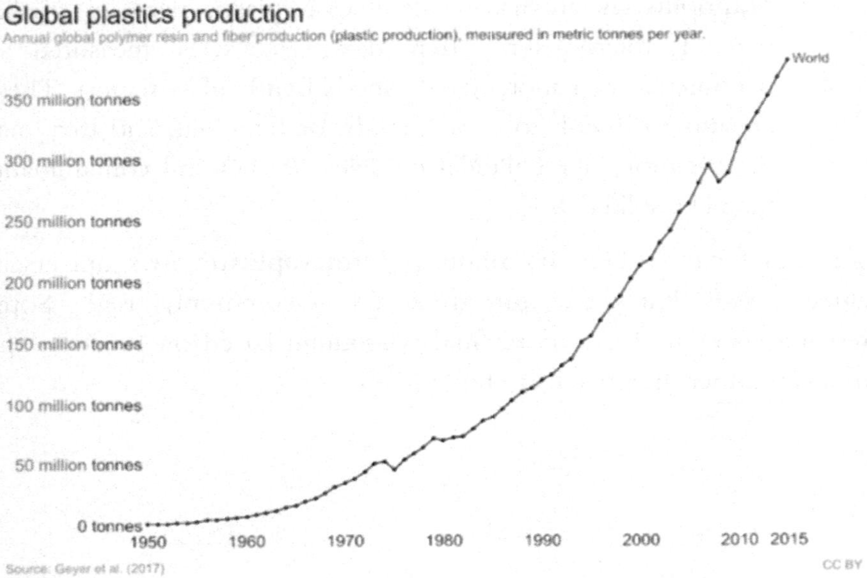

Figure 2: Global Plastics Production
Source: Our World In Data, CC BY 3.0 <https://creativecommons.org/licenses/by/3.0>, via Wikimedia Commons.

The size of a plastic particle significantly influences how it behaves in biological systems and ecosystems.

- **Microplastics** can be ingested by a wide variety of organisms, from plankton to whales, and may accumulate in the gastrointestinal tract or tissue.[7]
- **Nanoplastics**, due to their small size, can cross biological membranes, enter cells, interact with proteins and DNA, and potentially cause oxidative stress or inflammatory responses.[8]

Furthermore, smaller particles have a higher surface area-to-volume ratio, meaning they can adsorb more toxic chemicals such as heavy metals or persistent organic pollutants.[9] In effect, they can act as "Trojan horses," delivering toxins deep into organisms that ingest or inhale them.[10]

Composition and Common Polymers

Nano and microplastics are not a single type of material but a diverse group composed of various polymers. Common examples include:

- **Polyethylene** (PE): Found in plastic bags, bottles, and packaging films.
- **Polypropylene** (PP): Used in bottle caps, food containers, and textiles.
- **Polystyrene** (PS): Found in foam packaging, cups, and insulation materials.
- **Polyethylene terephthalate** (PET): Common in beverage bottles and synthetic clothing.
- **Polyvinyl chloride** (PVC): Used in plumbing, window frames, and flooring.
- **Acrylonitrile butadiene styrene** (ABS): Used in electronics and toys.[11]

Each polymer type differs in its durability, density, degradation rate, and potential for leaching additives. Many plastics contain additives like plasticizers, flame retardants, and UV stabilizers that may also pose health risks.[12]

Shape and Morphology

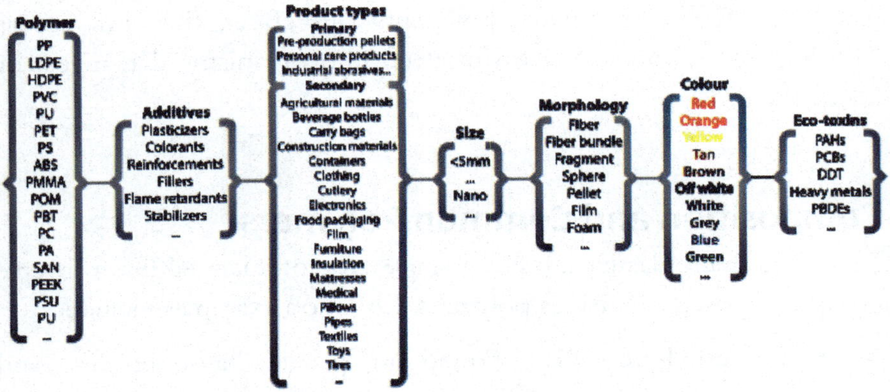

Figure 3: Shape & Morphology of Plastic
Source: Pidopram, CC BY-SA 4.0 <https://creativecommons.org/licenses/by-sa/4.0>, via Wikimedia Commons

Not all micro- and nanoplastics are shaped the same. Their physical form can affect how they interact with organisms and environmental systems.

- **Fragments**: Irregular, broken pieces from larger plastic items.
- **Fibers**: Thread-like shapes, often from clothing or ropes.
- **Beads**: Spherical shapes, often from cosmetics or industrial sources.
- **Films**: Thin, sheet-like particles from plastic bags or packaging.
- **Foams**: Lightweight, porous materials from packaging and insulation.[13]

Fibrous particles, for instance, are particularly common in indoor air due to textile shedding. These fibers are also the primary plastic pollutant found in laundry wastewater and airborne dust.[14]

Pathways of Formation

Most micro- and nanoplastics in the environment are secondary, formed through the degradation of larger plastic items. Degradation mechanisms include:

- **Photodegradation**: UV light from the sun breaks down plastic polymers, leading to brittleness and fragmentation.
- **Mechanical stress**: Wave action, abrasion, or industrial processes can physically break plastics into smaller pieces.
- **Thermal degradation**: High temperatures cause molecular breakdown.
- **Biodegradation**: Though limited, some microbes and enzymes can slowly degrade certain types of plastic under specific conditions.[15]

These degradation processes can take years or even centuries, depending on environmental conditions. In some marine environments, plastic bags have been found largely intact after 20 years.[16]

Analytical Challenges in Detection

Studying nano and microplastics is inherently difficult due to their small size and diversity. Detection methods include:

- **Microscopy**: Scanning electron microscopy (SEM) or fluorescence microscopy can visualize small particles.[17]
- **Spectroscopy**: Fourier-transform infrared (FTIR) and Raman spectroscopy identify polymer types based on their molecular vibrations.[18]

- **Mass spectrometry**: Used to analyze additives or leachates.[19]
- **Thermal analysis**: Techniques like pyrolysis–GC-MS can determine polymer composition by heating samples.[20]

However, nanoplastics often fall below the detection threshold of these instruments. Their small size, tendency to aggregate, and presence in complex matrices (e.g., sludge, tissue) make accurate quantification difficult.[21]

Moreover, contamination during sampling and analysis is a serious problem. Laboratories must wear cotton lab coats, use glass or metal equipment, and work in clean environments to avoid introducing airborne microfibers from the lab itself.[22]

Environmental Fate and Behavior

Once released into the environment, nano and micro-plastics follow complex pathways influenced by wind, water, density, and biological activity. Understanding these movements is essential for assessing exposure risks and designing remediation strategies.

A. Aquatic Systems

Rivers, lakes, and oceans are the most studied repositories of microplastics. Once introduced through wastewater discharge, stormwater runoff, or maritime activity, plastic particles distribute based on their buoyancy.

- **Buoyant plastics** like polyethylene and polypropylene tend to float and accumulate on the ocean surface, often ending up in gyres or "garbage patches".[23]

- **Denser polymers** such as PET and PVC sink, potentially disturbing benthic ecosystems.[24]
- **Microplastics** in the ocean are ingested by zooplankton, fish, crustaceans, and even whales.[25] In some cases, particles accumulate up the food chain, with top predators—and humans—bearing the brunt of contamination.[26]
- **Nanoplastics** may remain suspended in the water column, interacting with dissolved organic matter, trace metals, and other pollutants. These inter-actions alter their behavior and toxicity profile.[27]

B. Terrestrial Ecosystems

Although often overlooked, soils and sediments are major sinks for microplastics. Agricultural practices such as plastic mulch use, compost application, and irrigation with contaminated water contribute to the accumulation of plastic particles in farmland.[28]

Studies have detected microplastics in:

- **Compost and sewage sludge used as fertilizer.**[29]
- **Earthworm casts, indicating ingestion and egestion by soil fauna.**[30]
- **Groundwater sources, suggesting vertical migration through soil layers.**[31]

Plastic particles in soil can impair microbial activity, reduce fertility, and alter root development.[32] Moreover, nanoplastics may influence nutrient cycling by interacting with bacteria and fungi.[33]

C. Atmospheric Transport

Recent research has uncovered alarming evidence that microplastics are airborne. Tiny fibers and fragments become suspended in the air through:

- **Wear and tear of synthetic textiles.**[34]
- **Tire abrasion on roadways.**[35]
- **Dust from plastic waste and construction sites.**[36]

Wind and precipitation carry these particles over long distances. Snow samples from remote areas like the Pyrenees and Arctic have shown microplastic contami-nation, illustrating the atmospheric ubiquity of these materials.[37]

Indoor air often contains higher concentrations due to confined spaces, synthetic furniture, and limited ventilation. Inhalation of microfibers is now recognized as a significant exposure pathway, particularly in urban areas and occupational settings such as textile factories.[38]

Health Implications and Toxicology

Though research is ongoing, initial studies suggest that nano and microplastics can affect living organisms at multiple biological levels.

A. Physical Damage

Plastic particles can physically block or abrade internal organs. In marine life, microplastics have caused:

- **Gut blockage and false satiation in fish and birds.**[39]
- **Reduced feeding efficiency in plankton and filter feeders.**[40]

- **Internal injuries in crustaceans.**[41]

While human data is limited, ingestion of microplastics from seafood, salt, and water is already measurable[42]. There is concern that sharp or jagged particles could damage gastrointestinal linings.[43]

B. Chemical Leaching

Plastics contain numerous chemical additives, including:

- **Plasticizers** like phthalates.
- **Flame retardants** such as PBDEs.
- **Stabilizers, pigments, and bisphenol A** (BPA).

These additives can leach out and mimic hormones, disrupt metabolism, or interfere with developmental processes.[44] Phthalates and BPA, for instance, are classified as endocrine-disrupting chemicals (EDCs).[45]

Moreover, nano and microplastics can adsorb external pollutants like dioxins, PCBs, and heavy metals. When ingested, these particles act as vectors, delivering toxins deep into body tissues.[46]

C. Cellular and Molecular Effects

Nanoplastics, due to their small size, have been shown in lab settings to:

- **Penetrate cell membranes.**[47]
- **Interfere with mitochondrial function.**[48]
- **Trigger oxidative stress.**[49]
- **Promote inflammation.**[50]

Animal studies have linked nanoplastic exposure to immune suppression, altered gene expression, and reproductive toxicity.[51] Although human trials are ethically and practically challenging, cell culture and animal models provide compelling warning signs.[52]

Regulatory Definitions and Challenges

Governments and international organizations are beginning to recognize micro- and nanoplastics as a distinct pollutant category, but regulatory frameworks remain inconsistent.

- The European Union classifies microplastics as "intentionally added" if used in products, and is moving toward a ban on such uses.[53]
- The United States banned plastic microbeads in cosmetics in 2015, but has no broad legislation covering other sources.[54]
- Canada and Japan have developed monitoring programs, but lack enforceable limits on emissions.[55]

Key challenges include:

- Absence of standardized measurement tech-niques.[56]
- Disagreement on size thresholds and health benchmarks.
- Legal ambiguity over plastic additives and secondary degradation.[57]

As scientific evidence mounts, pressure is growing on regulatory bodies to define, monitor, and restrict these pollutants systematically.[58]

Public Awareness and Media Influence

Mass media has played a crucial role in raising awareness about nano and microplastics. Documentaries like *Blue Planet II* and *A Plastic Ocean* have brought the issue to global audiences.[59]

Viral images of plastic-strewn beaches and marine animals entangled in debris have fueled public outrage and consumer shifts.[60]

Social media campaigns like **#PlasticFreeJuly** and **#BreakFreeFromPlastic** have engaged millions of people worldwide.[61] While awareness alone doesn't solve the crisis, it generates pressure on governments and corporations to change policies and supply chains.

Educational institutions are also incorporating plastic pollution into science and environmental curricula, ensuring the next generation of citizens is better informed than their predecessors.[62]

Innovations in Detection and Analysis

As scientific concern around micro- and nanoplastics grows, so does the development of tools to detect and study them. Traditional methods like light microscopy and FTIR spectroscopy are useful for identifying larger microplastics, but they fall short with particles at the nanoscale. Cutting-edge innovations now allow for greater resolution, specificity, and throughput in particle detection and chemical characterization.

A. Spectroscopy Advancements

- **Raman microscopy** offers high spatial resolution and can identify even small fragments based on vibrational fingerprints of polymer chains.[63]

- **Near-infrared spectroscopy** (NIR) allows rapid scanning of materials and is increasingly being adapted for environmental samples.[64]
- **FTIR imaging combined with automated software** can now analyze thousands of particles per sample in a semi-automated way, speeding up large-scale environmental surveys.[65]

B. Thermal and Mass Spectrometry

- **Pyrolysis–gas chromatography–mass spectrometry** (Py-GC-MS) breaks down plastics thermally and identifies the resulting compounds, allowing for precise polymer identification.[66]
- **Thermogravimetric analysis** (TGA) measures mass loss as a function of temperature, useful for studying degradation and estimating plastic content in mixed samples.[67]

C. Nanoplastic Detection

This remains a major challenge due to size, aggregation behavior, and low environmental concentrations. Emerging approaches include:

- **Nanoparticle tracking analysis** (NTA) to estimate size distribution and particle count.[68]
- **Dynamic light scattering** (DLS) for assessing nanoscale colloids.[69]
- **Flow cytometry** adapted for synthetic polymers, although signal interference remains an obstacle.[70]

Researchers are now also exploring biosensors, fluorescent tagging, and atomic force microscopy (AFM) for nanoplastic detection in complex biological tissues.[71]

Interaction with Other Pollutants

Nano and microplastics rarely exist in isolation. In the environment, they encounter a cocktail of pollutants—metals, hydrocarbons, pesticides—that can alter their behavior, toxicity, and transport.

- **Adsorption of heavy metals** like lead, cadmium, and mercury onto plastic surfaces increases their toxicity.[72]
- **Persistent organic pollutants** (POPs) such as PCBs and dioxins bind readily to plastic due to their hydrophobic nature.[73]
- **Microbial colonization** forms "plastispheres" where bacteria, including pathogens and anti-biotic-resistant strains, thrive.[74]

This interaction turns plastic particles into complex, dynamic vehicles of chemical and biological contamination, with effects that are difficult to predict or model.[75]

The Circular Economy and Design Solutions

While much of the discussion around microplastics focuses on remediation, a growing movement within industry and policy circles is pushing for circular economy principles. The idea is to prevent plastic pollution by redesigning systems of production and consumption so that waste is minimized or eliminated.[76]

A. Eco-Design of Products

Companies are now experimenting with:

• **Biodegradable polymers** that break down in natural environments (e.g., PLA, PHA).[77]

• **Edible packaging** made from algae or starch-based materials.[78]

• **Modular design** that allows products to be disassembled and recycled more easily.[79]

B. Extended Producer Responsibility (EPR)

This policy approach shifts the burden of waste management from consumers and municipalities to the companies that produce plastic products.[80] EPR schemes can include:

• **Mandatory take-back programs.**

• **Deposit-refund systems.**

• **Penalties for non-recyclable packaging.**[81]

Such policies incentivize manufacturers to reduce plastic use, improve recyclability, and invest in alternative materials.[82]

C. Closed-Loop Recycling

Advanced recycling technologies like chemical depolymerization aim to break plastics down to their monomers so they can be repolymerized into new products without loss of quality.[83] Although energy-intensive, these technologies promise a more sustainable future than mechanical recycling alone.[84]

Ethical and Philosophical Considerations

Nano and microplastic pollution isn't just a technical or environmental issue—it's an ethical one. It raises questions about responsibility, equity, and inter-generational justice.[85]

- Who is responsible for the harm caused by plastic pollution? Is it the individual consumer, the producer, or the regulatory system that allows it?
- Who suffers most from plastic exposure? Often, it's the poor, marginalized communities, or developing nations where waste is exported.[86]
- What do we owe to future generations who will inherit an ecosystem saturated with synthetic particles?[87]

The pervasive nature of nano and microplastics calls for a rethinking of our material culture. Plastic has enabled enormous technological progress, but it has also tied humanity into a web of persistent and invisible pollution.[88]

This dilemma challenges us to consider not only better technology or stricter policy but a fundamental shift in values: toward sustainability, humility, and long-term thinking.[89]

Final Thoughts

Nano and microplastics are more than just environmental debris—they are symbols of a broken system of production and consumption. They embody the unintended consequences of technological progress, the hidden costs of convenience, and the silent spread of synthetic material through every corner of the biosphere.[90]

This chapter has explored the definitions, chemistry, environmental behavior, health implications, detection methods, and ethical considerations of these particles. It has shown that size matters—not only in determining how these materials behave but also in how we perceive and manage them.

To tackle this challenge, we need a multi-pronged approach:

- Scientific innovation to improve detection and assess toxicity.[91]
- Regulatory clarity and consistency across regions.
- Cultural shifts in how we use, value, and dispose of materials.[92]

The chapters ahead will build on this foundation, exploring how plastic pollution emerged, where it resides, what harms it causes, and—most importantly—what can be done to address it.

In an age defined by global complexity, the issue of nano and microplastics teaches a profound lesson: the smallest things can have the biggest impacts.[93]

Chapter 2: The Rise of Plastic Pollution

Introduction: From Marvel to Menace

Plastic was once hailed as a miracle material—a symbol of innovation, affordability, and modern convenience. Its rise in the 20th century paralleled breakthroughs in chemistry, industry, and post-war consumerism.[1] For a time, plastics seemed to offer limitless possibilities: lightweight, durable, moldable, and cheap. But as its use exploded, so did its environmental footprint.[2] The very qualities that made plastic so desirable—resilience, chemical resistance, permanence —became liabilities.[3]

This chapter chronicles the evolution of plastic from its invention to its mass production and distribution, tracing how nano and microplastics emerged as unintended consequences. Understanding this history is key to contextualizing the modern crisis and recognizing the systemic roots of plastic pollution.[4]

The Early Origins of Synthetic Polymers

The story of plastics begins in the 19th century with the search for alternatives to scarce natural materials like ivory and shellac.[5]

- 1869: John Wesley Hyatt invents celluloid, the first semi-synthetic plastic, from cellulose treated with camphor.[6] It was marketed as an ivory substitute for billiard balls, combs, and piano keys.

- 1907: Leo Baekeland develops Bakelite, the first fully synthetic plastic made from phenol and formaldehyde.[7] It was heat-resistant and insulating, ideal for electrical applications and household items.

These early plastics were revolutionary but limited in scope. They were also expensive and somewhat brittle. The true explosion came with World War II.[8]

Wartime Innovation and Industrialization

The global conflicts of the 20th century catalyzed plastic innovation. Metals, rubber, and silk were diverted to war efforts, and the military needed affordable substitutes.[9]

- Nylon, developed by DuPont in 1935, replaced silk in parachutes, ropes, and stockings.[10]
- Polyethylene, first synthesized in the 1930s, was mass-produced for radar cable insulation.[11]
- Polystyrene, previously used in packaging, was adapted for military use in flotation devices and food containers.[12]

After the war, the infrastructure for plastic production was already in place. Industries pivoted from tanks and weapons to consumer goods, and plastics became the backbone of post-war prosperity.[13]

The Boom of Consumer Plastics

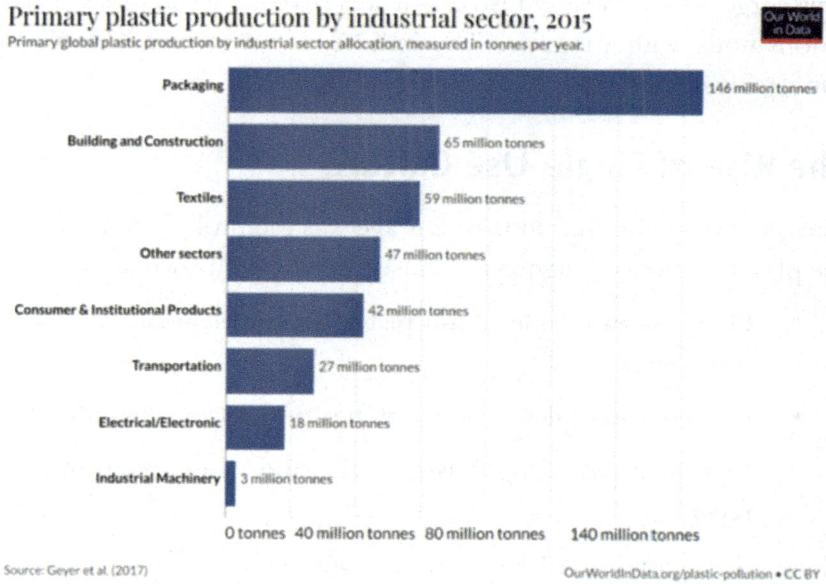

Figure 4: Primary Plastic Production (Industrial Sector)
Source: Our World in Data, CC BY 4.0 <https://creativecommons.org/licenses/by/4.0>, via Wikimedia Commons.

From the 1950s to 1970s, plastic consumption exploded. Advertisements promoted plastic as a hygienic, space-age material that liberated households from drudgery.[14]

- Tupperware became a symbol of domestic modernity.
- Plastic bags replaced paper in grocery stores.
- Vinyl and polyester entered fashion and upholstery.
- PET (polyethylene terephthalate) revolutionized beverage containers.[15]

Annual global plastic production skyrocketed from 1.5 million metric tons in 1950 to over 50 million by 1976.[16]

This surge was not just industrial—it was psychological. Plastic became synonymous with progress. The culture of convenience took root, reinforcing disposability as a norm.[17]

The Rise of Single-Use Culture

The most dramatic shift in plastic usage was the proliferation of single-use plastics—items designed to be used once and thrown away.[18]

- Plastic straws, cutlery, and plates became standard at fast-food restaurants.
- Bottled water sales soared with health-conscious branding.
- Clamshell packaging, blister packs, and shrink wrap took over retail.

These items, often used for minutes, were made from materials that persist for centuries.[19] Landfills filled up, and litter began to spread across roadsides, beaches, and waterways. But at the time, the environmental cost was largely invisible or ignored.[20]

Early Environmental Warnings

By the late 1960s and 1970s, the honeymoon period with plastics began to fray.[21] Marine scientists started to document unusual debris on coastlines. In 1972, a landmark paper in *Science* reported plastic particles in the North Atlantic.[22]

Other early observations included:

- Entanglement of seabirds in six-pack rings.
- Fish with plastic fragments in their guts.
- Oceanographers noticing persistent litter far offshore.[23]

These findings were met with concern, but not decisive action. Instead, industry groups launched anti-litter campaigns, shifting blame to consumers rather than addressing production levels.[24] One of the most famous slogans, "People Start Pollution, People Can Stop It," emerged from a coordinated strategy to avoid regulatory backlash.[25]

Microplastics: The Crisis Within a Crisis

Microplastics—plastic particles less than 5mm—were first observed in the 1970s but didn't enter mainstream discourse until the 2000s.[26] Their sources are diverse:

- Primary microplastics: intentionally manufactured (e.g., exfoliating beads, industrial abrasives).
- Secondary microplastics: formed from the fragmentation of larger plastic waste through UV radiation, mechanical abrasion, or biological processes.[27]

These particles permeate every environment—oceans, rivers, soil, and air. In the last decade, researchers have found microplastics in:

- Sea salt.[28]
- Bottled and tap water.
- Beer.
- Honey.
- Human stool samples.[29]

The scale of contamination is staggering. A 2019 study estimated that the average American ingests over 70,000 microplastic particles annually.[30]

The Birth of Nanoplastics

While microplastics have received increasing attention over the past two decades, nanoplastics—particles smaller than 1 micrometer—are a more recent and concerning discovery. These particles often originate from the further degradation of microplastics, tire abrasion, synthetic textiles, or cosmetic products, and they pose unique risks because of their extreme smallness, mobility, and ability to cross biological barriers.

Nanoplastics are particularly worrisome for several reasons:

- They can enter human cells, passing through membranes and potentially disrupting cellular function.[31]
- They are undetectable by the naked eye, meaning their presence in water, air, and food is easily overlooked.[32]
- They exhibit high surface-area-to-volume ratios, which can increase their ability to carry toxic substances.[33]

The understanding of nanoplastics is still developing, but early studies show they bioaccumulate and may act as endocrine disruptors or oxidative stress inducers.[34] These findings compound the urgency of tackling plastic pollution at all levels—macro, micro, and nano.

Global Plastic Production: An Unstoppable Climb

The numbers behind plastic production help illustrate the scale of the crisis. According to the United Nations Environment Programme (UNEP) and OECD:

- 1950: 1.5 million metric tons produced globally.[35]
- 1970: 35 million metric tons.
- 1990: 100 million metric tons.
- 2010: 270 million metric tons.
- 2023: Over 400 million metric tons annually.[36]

That's more than 50 kilograms of plastic produced per person every year, globally. Much of this growth comes from developing economies experiencing rising consumer demand and lacking infrastructure to manage waste responsibly.[37]

Plastic production is expected to double again by 2040, driven by packaging, textiles, and construction materials.[38] Without intervention, this trajectory will only deepen the crisis.

Global Waste Management Failures

Despite decades of recycling campaigns, the world has largely failed to establish an effective circular economy for plastics. According to 2022 figures:

- 9% of all plastic ever produced has been recycled.[39]
- 12% has been incinerated.

- 79% has accumulated in landfills, dumps, or the natural environment.[40]

Several factors contribute to this failure:

- Downcycling: Most plastic is not truly recycled into new products but degraded into lower-quality materials.[41]
- Contamination: Mixed materials and food residue make recycling inefficient and expensive.
- Market economics: Virgin plastic is often cheaper than recycled plastic, especially when oil prices are low.
- Exportation of waste: Wealthy nations have historically exported plastic waste to developing countries with poor regulatory oversight.[42]

The result? Mountains of unprocessed waste, much of which breaks down into micro- and nanoplastics over time.

Ocean Gyres and the Plastic Soup

One of the most visible consequences of plastic pollution is the formation of ocean gyres—massive rotating currents where floating plastic debris accumulates. The most infamous is the Great Pacific Garbage Patch, which spans a region estimated at 1.6 million square kilometers, roughly twice the size of Texas.[43]

Contrary to popular belief, these are not solid "islands" of trash. They are more like plastic soups—an ever-moving blend of macroplastics, microplastics, fishing nets, and fragments suspended in the water column.[44] These gyres exist in all major oceans:

- North Atlantic Gyre.

- South Atlantic Gyre.
- Indian Ocean Gyre.
- North Pacific Gyre.
- South Pacific Gyre.

Oceanic plastic doesn't remain static. Waves, currents, and animal ingestion move it across vast distances, making national borders irrelevant. A plastic bottle discarded in California could wash up in the Philippines, or end up as microplastic in Arctic ice.[45]

Human Footprint in Remote Locations

One of the most shocking aspects of plastic pollution is how it affects even the most remote parts of Earth. Scientists have found microplastics in:

- Arctic sea ice.
- Mountain snow in the Alps and Himalayas.
- Deep-sea sediments.
- Remote desert regions.
- Freshwater lakes and groundwater.[46]

A 2020 study published in *Nature Geoscience* reported that microplastic particles can be transported through the atmosphere and deposited by rainfall—suggesting a plastic "water cycle" of sorts.[47] This aerial dispersal mechanism explains the presence of plastic particles thousands of miles from any human settlement or manufacturing source.

Such findings challenge the notion that plastic pollution is a localized issue. It is clearly a global phenomenon, one that affects every biome on Earth—from the depths of the Mariana Trench to the peaks of Everest.

Plastic and Wildlife: An Ecological Disaster

Plastics are mistaken for food by hundreds of species:

- Sea turtles consume plastic bags, mistaking them for jellyfish.[48]
- Seabirds are found dead with stomachs full of bottle caps, lighters, and packaging fragments.[49]
- Whales have washed ashore with entire car bumpers and fishing nets in their stomachs.[50]

The result is a combination of:

- Starvation: Animals feel full but receive no nutrients.
- Internal injuries: Sharp fragments can pierce organs.
- Reproductive harm: Exposure to endocrine-disrupting compounds can alter hormone function.[51]

And it's not just marine life. On land, microplastics affect soil microbes, earthworms, and insects, disrupting nutrient cycling and pollination. Studies also show microplastics in cattle feed, affecting livestock health and entering the human food chain.[52]

This environmental toll is not merely collateral damage—it is structural, predictable, and deeply interconnected with industrial behavior.

The Invisible Cost: Health and Economics

As plastic pollution spread, it introduced unseen costs to human health and global economies. While marine wildlife and picturesque beaches showed the visible signs of degradation, the human toll was unfolding in laboratories, hospitals, and policy centers.[53]

Microplastics and nanoplastics have now been found in:

- Human lungs.
- Placenta tissue.
- Breast milk.
- Bloodstreams.[54]

These particles don't just pass through the body—they may accumulate, interact with immune cells, and carry toxic chemicals like bisphenol A (BPA), phthalates, and flame retardants. These additives are linked to:

- Hormone disruption.
- Lower sperm counts.
- Neurological issues.
- Autoimmune responses.
- Increased cancer risk.[55]

Healthcare systems will likely bear the future burden of chronic diseases associated with long-term, low-dose exposure to synthetic compounds. These links are still under investigation, but the precautionary principle suggests action is warranted even without definitive causation.[56]

Economically, the UNEP estimates that marine plastic pollution alone costs the global economy $13 billion annually, including losses to:

- Fisheries.
- Coastal tourism.
- Clean-up operations.
- Ecosystem services.[57]

These figures don't include the costs of inland pollution, waste management failures, or health consequences. When considered holistically, the true economic footprint of plastic is vastly underestimated.[58]

Industry Push-back and Greenwashing

Efforts to regulate or ban plastic items have been met with significant **push-back from industry lobbies**, especially from petrochemical and packaging giants. These corporations invest heavily in public relations campaigns that:

- Promote individual responsibility (e.g., "Don't litter" messaging).
- Emphasize **recycling** while lobbying against deposit schemes.
- Fund studies downplaying plastic toxicity.[59]

A phenomenon known as **greenwashing**—marketing products as "eco-friendly" or "biodegradable" without meaningful change—has become widespread. Bioplastics, for instance, may only biodegrade under industrial composting conditions, which are rare in most countries. These misleading practices create confusion, stall progress, and undermine legitimate environmental efforts.[60]

The Recycling Myth

For decades, consumers were encouraged to believe that their plastic waste was being efficiently recycled. However, the truth is far less reassuring:

- Most plastics are not recyclable in practice, especially those contaminated by food or made from composite materials.[61]
- Even recyclable plastics can only go through the process once or twice before degrading in quality.[62]
- The iconic recycling symbol (♻) with a number inside doesn't mean the item is being recycled—it denotes the type of plastic, many of which lack local processing capacity.[63]

In 2018, China implemented the National Sword policy, which banned the import of contaminated plastic waste.[64] This triggered a global reckoning as Western countries scrambled to find alternatives—many turning to Southeast Asia, where lack of oversight led to environmental injustice.[65]

The world is now left with a deluge of plastic waste and no sustainable way to manage it.

Policy Milestones and Missed Opportunities

Governments have taken some steps to address plastic pollution:

- EU Single-Use Plastics Directive (2019): Banned plastic straws, stirrers, and cutlery.[66]
- Canada's 2022 ban on certain single-use plastics.[67]

- Kenya's plastic bag ban (2017): One of the strictest in the world.[68]
- California and Oregon: Introduced extended producer responsibility (EPR) laws.[69]

These initiatives are important—but often piecemeal and inconsistently enforced. A lack of global coordination, weak penalties, and industry influence weaken their impact. Voluntary agreements and pledges rarely translate into action.[70]

In 2022, the United Nations Environment Assembly (UNEA) passed a historic resolution to create a legally binding treaty on plastic pollution by 2024.[71] While this is promising, negotiations are ongoing, and enforcement mechanisms remain uncertain.

The Cultural Shift: From Convenience to Consequence

The tide is beginning to turn culturally. Documentaries like *A Plastic Ocean*, *The Story of Plastic*, and *Seaspiracy* have brought the issue into public consciousness.[72] Social media movements such as **#PlasticFreeJuly**, **#BreakFreeFromPlastic**, and **#ZeroWaste** are influencing consumer behavior.[73]

Some notable changes include:

- A rise in refill shops and bulk stores.
- Adoption of solid shampoo bars, reusable wraps, and glass containers.
- An emerging class of eco-conscious influencers driving product reviews and lifestyle shifts.

Youth-led movements, especially those connected to climate justice campaigns, have made plastic pollution a moral issue as much as an environmental one.[74]

However, systemic change is still needed. Individual actions matter, but they must be matched by top-down regulatory reform and corporate accountability to truly curb plastic's dominance.[75]

Looking Ahead: Tipping Points and Innovations

The urgency of the plastic crisis is driving innovation across multiple sectors. Researchers and startups are exploring:

- Enzymes and microbes that can break down PET and polystyrene.[76]
- Plastic alternatives made from algae, mushrooms, seaweed, and agricultural waste.[77]
- Blockchain-based tracking systems for supply chain transparency.[78]
- Reusable packaging ecosystems, such as TerraCycle's Loop system.[79]

At the same time, educational institutions are adding plastic literacy to environmental curricula, and citizen science projects are equipping communities to monitor plastic in their local ecosystems.[80]

A true solution will require a multidimensional strategy:

- Cap production of virgin plastics.
- Design for reuse and repair.

- Penalize greenwashing.
- Support circular economies.
- Center environmental justice.

The transition away from plastic dominance will not be easy, but neither is it optional. If we fail to act, by 2050 there could be more plastic than fish in the ocean by weight [81]—a haunting symbol of misaligned priorities.

Conclusion: From Wonder Material to Global Threat

Plastic began as a triumph of science—a moldable marvel that revolutionized human life. But its success story became a cautionary tale of overproduction, poor foresight, and corporate negligence. From 19th-century billiard balls to 21st-century nanoplastics in baby formula, its arc traces the complex interplay between innovation, capitalism, and environmental degradation.[82]

Understanding the rise of plastic pollution is not just a matter of history—it is a foundation for action. If we know how we got here, we can change where we're going. The next chapter will explore just how widespread plastics have become across Earth's ecosystems and food chains, setting the stage for deeper analysis of their origins and impacts.

Chapter 3: Where Plastics Lurk: A Global Ubiquity

Introduction: A World Awash in Plastic Particles

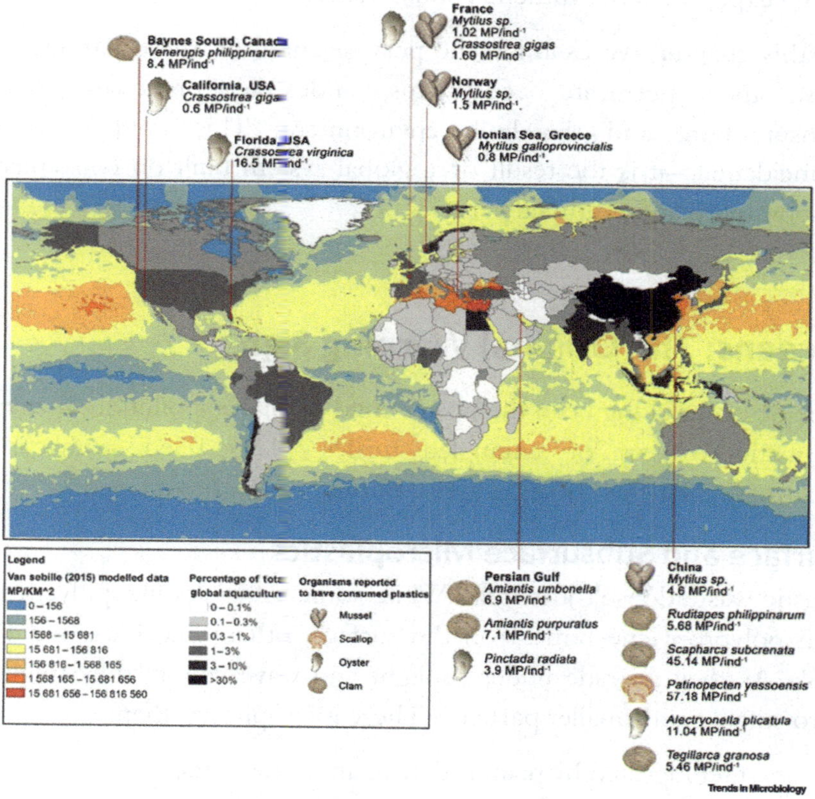

Figure 5: Worldwide Plastic Particle accumulation
Source: Jake Bowley, Craig Baker-Austin, Adam Porter, Rachel Hartnell and Ceri LewisArcGIS version 10.6.1, CC BY-SA 4.0 <https://creativecommons.org/licenses/by-sa/4.0>, via Wikimedia Commons

When we think of plastic pollution, we often imagine floating debris in the ocean or litter strewn across beaches. But plastic pollution today is far more insidious. Tiny plastic particles, often invisible to the naked eye, are now found everywhere—embedded in soil, drifting through the atmosphere, circulating in oceans, and infiltrating the human body. Microplastics (less than 5mm) and nanoplastics (less than 1µm) have colonized nearly every environment on Earth, turning our planet into a living experiment in synthetic contamination.[1]

In this chapter, we examine the pervasiveness of these particles and how they permeate ecosystems, industrial processes, human consumption, and the built environment. This ubiquity is not coincidental—it is the result of a global system built on convenience, disposability, and synthetic permanence.

Oceans: The Original Warning Sign

The ocean has long been the poster child of plastic pollution, thanks to dramatic imagery of entangled turtles, plastic-filled stomachs of albatrosses, and vast floating gyres. But the crisis goes deeper.

Surface and Subsurface Microplastics

Plastic waste doesn't just float. While some polymers like polyethylene and polypropylene remain on the surface, others like PET and PVC sink. As they degrade under sunlight and wave action, they fragment into smaller and smaller particles. These microplastics then:

- Get ingested by plankton, fish, and filter feeders.
- Sink into deep-sea sediments.
- Travel vast distances through ocean currents.

Even the remote **Mariana Trench**, the deepest point on Earth, has recorded the presence of synthetic fibers and fragments.[2]

Gyres and Garbage Patches

The infamous **Great Pacific Garbage Patch** between Hawaii and California is not a visible island of trash, but a diffuse soup of microplastics suspended in water. Similar zones exist in all five subtropical oceanic gyres, with studies estimating **over 5 trillion plastic pieces** currently floating in the world's oceans.[3]

Marine Food Chain Contamination

Shellfish, small fish, and even whales ingest microplastics, mistaking them for food. These particles bioaccumulate and **biomagnify** as they move up the food chain, eventually reaching human consumers.[4] For coastal communities that rely on seafood, this is not an abstract concern—it's a direct threat.

Freshwater and Drinking Water

While oceans have captured the spotlight, **freshwater systems**—rivers, lakes, reservoirs, and even groundwater—are just as affected.

Tap and Bottled Water

In a 2018 study conducted by Orb Media and the State University of New York, **93% of bottled water samples** from major brands contained microplastic contamination.[5] Shockingly, tap water samples also tested positive, suggesting that **no source of drinking water is immune**.

Sources of Contamination

Plastics enter freshwater through:

- Runoff from urban environments.
- Tire wear particles washed into storm drains.
- Wastewater treatment plant discharge.
- Degraded agricultural mulch films and irrigation systems.

The **Danube River** in Europe, for example, transports an estimated **1,500 tons of plastic** into the Black Sea each year.[6] Major rivers like the Ganges, Yangtze, and Amazon act as superhighways for inland plastic to reach marine ecosystems.

Airborne Plastics: The Inhalable Threat

Few people consider that the air they breathe could contain plastic, but **airborne microplastics** are now a verified and troubling vector.

Indoor Environments

Inside homes and offices, plastic particles originate from:

- Synthetic textiles (e.g., polyester, acrylic).
- Upholstery and carpets.
- Household dust laden with microfibers.

Studies have found that **indoor air contains up to five times more microplastics** than outdoor air.[7] Central heating, vacuuming, and simple movements stir particles into suspension. Infants crawling on carpets inhale and ingest higher volumes relative to their body weight.

Outdoor Dispersal

Even remote areas are not safe. Plastic particles have been found in:

- Arctic snow samples.

- Mountain air over the Pyrenees.
- National parks in the U.S.

These particles are carried via **atmospheric deposition**—lifted by wind and turbulence, carried by thermal currents, and deposited by precipitation.[8]

Respiratory Health Concerns

Inhalation of microplastics can introduce particles into the:

- Nasal passages.
- Bronchial tubes.
- Alveolar sacs.

Laboratory studies suggest that such exposure may lead to inflammation, oxidative stress, and possibly long-term respiratory issues.[9] Nanoplastics, due to their size, may even cross into the bloodstream via lung tissue.[10]

Soil and Agriculture

It's easy to assume that plastic particles remain on the surface, but they also infiltrate **terrestrial environments**, especially soil systems. Agricultural practices, ironically meant to sustain life, contribute heavily.

Sources in Soil

- Degraded plastic mulch used in farming.
- Sewage sludge applied as fertilizer (biosolids).
- Irrigation with contaminated water.

- Litter breakdown in pastures and fields.[11]

Soil Ecosystem Disruption

Microplastics in soil have been shown to:

- Alter microbial communities.
- Reduce earthworm reproductive rates.
- Affect root development in some plants.[12]

This is especially concerning because **soil health underpins global food security**. What begins as contamination in agricultural land may ultimately impact crop yield, nutritional content, and plant-microbe relationships.

Food and Beverages

Plastic particles do not just surround our food—they are **in our food**. Recent studies have demonstrated that microplastics and nanoplastics can be present in many common food items, either through direct contamination or through bioaccumulation in the food chain.

Seafood

Shellfish and small fish are especially prone to ingesting microplastics due to their filter-feeding behavior. Mussels, oysters, and anchovies have been repeatedly shown to contain synthetic particles in their tissues. Consumers eating these animals whole—including the digestive tract—are directly consuming plastic.[13]

Salt

Table salt has emerged as a consistent source of microplastic exposure. Samples from over a dozen countries found particles in commercial sea salts, lake salts, and rock salts. Since the salts are derived from natural sources like evaporated seawater, the **ubiquity of oceanic microplastics guarantees contamination.**[14]

Honey and Sugar

Even highly processed foods are not immune. Studies from Germany and the U.K. have found that **honey and granulated sugar** contain microplastic fragments, likely introduced during processing or packaging.[15] Many production lines still rely on plastic containers, pipes, and wrap materials that shed particles over time.

Fruits and Vegetables

Fruits and vegetables can also carry plastics, especially those grown in contaminated soil or irrigated with polluted water. Nanoplastics may penetrate root systems and be absorbed into plant tissues, although this is an emerging field of study.[16] Meanwhile, surface contamination during washing, packaging, and transport remains a consistent vector.

Processed and Packaged Foods

Perhaps the most plastic-saturated items in modern diets are **processed and prepackaged foods**. These include:

- Ready-made meals in plastic trays.
- Bottled drinks.
- Snack foods in multi-layer plastic bags.

Leaching of chemicals such as **bisphenol A (BPA)** and **phthalates** occurs especially when food is heated in plastic, frozen in plastic, or stored over time.[17] These additives are endocrine disruptors and may contribute to a range of metabolic and reproductive disorders.

Cosmetics and Personal Care Products

Plastics aren't limited to what we eat—they're also found in what we **apply to our skin**. Microbeads, once common in exfoliants and toothpaste, are a well-known example, but microplastics persist in many modern formulations.

Common Plastic Ingredients

- **Polyethylene (PE).**
- **Polypropylene (PP).**
- **Polymethyl methacrylate (PMMA).**
- **Nylon-12.**[18]

These ingredients serve as binding agents, fillers, and texture enhancers in everything from moisturizers to foundations. Although some countries have banned microbeads, loopholes often allow **liquid or soluble plastic polymers** to remain.

Routes of Exposure

- **Dermal Absorption**: Small enough molecules may penetrate skin layers.
- **Inhalation**: Powders, sprays, and aerosols release plastic particles into breathable air.
- **Ingestion**: Lip products and hand-to-mouth contact facilitate transfer into the digestive tract.[19]

Even if plastics in personal care products don't immediately harm the individual, they **rinse down drains** into sewage systems and eventually re-enter the environment.

Textiles and Clothing

A lesser-known but massive source of microplastic pollution is the **clothing industry**, particularly synthetic textiles like:

- **Polyester.**
- **Acrylic.**
- **Nylon.**
- **Spandex.**[20]

When these materials are worn, washed, or dried, they shed **microfibers**—tiny threads of plastic that enter the air and water. A single load of laundry can release hundreds of thousands of microfibers, which pass through wastewater treatment plants and end up in rivers and oceans.

Dryers and Airborne Spread

Tumble dryers, particularly those without adequate filtration, emit synthetic fibers directly into the air. These fibers settle on surfaces, infiltrate the respiratory system, and are even found on food plates before meals are served.[21]

The Fast Fashion Dilemma

Fast fashion—characterized by cheaply made, disposable garments—uses more synthetic fabrics, contributing to a higher microfiber shedding rate. These clothes deteriorate quickly, ending up in landfills or incinerators, where they release even more pollutants.[22]

Indoor Environments: A Plastic Microcosm

Home and office interiors are increasingly recognized as **microplastic hotspots**, often more contaminated than outdoor settings.

Sources Include:

- Upholstered furniture and cushions.
- Wall-to-wall carpets made from nylon or polyester.
- Vinyl flooring.
- Plastic-coated electronics.
- Plastic storage bins and organizers.[23]

Dust as a Carrier

Indoor dust is a cocktail of skin flakes, pollen, household chemicals—and microplastics. Children, especially toddlers who play on the floor and put objects in their mouths, have a **higher ingestion rate** than adults. Studies estimate that children may ingest tens of thousands of microplastic particles annually from indoor dust alone.[24]

Ventilation and Filters

Most heating, ventilation, and air conditioning (HVAC) systems are not designed to filter out plastic particles. Without HEPA-level filtration, microplastics continue to circulate within buildings, creating chronic low-level exposure.[25]

Human Tissues and Fluids

In recent years, breakthrough studies have confirmed what many suspected: microplastics and nanoplastics are now present **within the human body**.

Findings Include:

- **Blood**: Detected in ~80% of sampled individuals (Netherlands, 2022).
- **Lungs**: Found in both upper and lower respiratory tissues.
- **Placenta**: Microplastics detected in all layers, suggesting fetal exposure.
- **Breast Milk**: Synthetic polymers detected in samples taken from mothers in Italy.[26]

These findings raise significant concerns about **bioaccumulation**, toxicity, and cellular interactions. Because nanoplastics are small enough to pass through biological membranes, they may cross the **blood-brain barrier**, the **placental barrier**, and even penetrate into organ tissues.[27]

While long-term health outcomes are still being studied, early models suggest potential links to:

- Inflammation.
- Oxidative stress.
- Endocrine disruption.
- Neurodevelopmental issues.[28]

Microplastics in the Food Chain: From Soil to Table

The infiltration of microplastics into food systems isn't just direct—it also happens indirectly through the **bioaccumulation and biomagnification processes** within ecosystems. When small organisms ingest plastic particles, those plastics travel up the food web, ultimately reaching top predators—humans included.

Terrestrial Food Chains

- **Livestock exposure**: Grazing animals may ingest plastic through contaminated grass, soil, or water. Ingested microplastics may remain in the gut or enter tissues, although more research is needed.
- **Manure contamination**: When animal waste is used as fertilizer, it may introduce plastic particles back into agricultural fields, forming a **closed loop of exposure**.[29]

Aquatic Food Chains

Fish and shellfish consume microplastics directly or through prey, accumulating particles in their digestive systems and tissues. Larger marine predators and birds often suffer severe blockages, malnutrition, or poisoning due to plastic ingestion.

Researchers have documented microplastics in **mollusks, crustaceans, plankton, and even sea cucumbers**—organisms at all levels of the aquatic food chain. Humans consuming seafood become the **apex recipients** of this synthetic load.[30]

Plastic-Contaminated Ecosystems: Case Studies

1. Arctic and Antarctic

Polar regions were once assumed to be pristine, but microplastics have now been found in **Arctic sea ice, Antarctic snow, and polar animal feces**. These discoveries underscore the **global transport mechanisms**—air currents, ocean flows, and migratory species.[31]

2. The Pyrenees Mountains

A 2019 study recorded microplastics in remote mountaintop regions of the Pyrenees, deposited via atmospheric transport. This means microplastic particles can travel **over 100 kilometers** from the nearest urban source.[32]

3. Urban Environments

Cities like Paris and London have documented **airborne microplastic fallout** on building surfaces, cars, and open food. Urban areas are thought to be microplastic "hotspots" due to dense traffic, industrial emissions, and synthetic building materials.[33]

Waste Management as a Distribution Channel

Ironically, waste management systems intended to mitigate plastic pollution often serve as unintentional **distribution networks** for microplastics.

Recycling Facilities

Recycling plants grind, melt, and shred plastic waste. These mechanical processes can release large quantities of airborne microplastics if not properly ventilated. Moreover, most facilities are not equipped to handle **microplastic residue**—only bulk items.[34]

Landfills

In landfills, plastic degrades slowly due to UV exposure, moisture, and microbial activity. Leachate—the liquid runoff from landfills—can carry microplastics into groundwater and nearby waterways.[35]

Incineration

Incinerating plastic does not eliminate the problem—it simply changes its form. Burning plastics can produce **nanoparticles** that escape filtration systems and disperse into the air. Incinerators also generate toxic ash and emissions that may include **persistent organic pollutants (POPs)**.[36]

The Illusion of "Biodegradable" Plastics

A growing market exists for so-called **biodegradable plastics**, marketed as environmentally friendly alternatives. However, many of these products do not degrade efficiently in natural settings.

PLA and Oxo-Degradable Plastics

- **Polylactic acid (PLA)** is often used in compostable cutlery and cups but requires **industrial composting facilities** (high heat, controlled humidity) to break down.
- **Oxo-degradable plastics** fragment into microplastics faster but do not truly degrade—they simply **accelerate the formation of microplastics**.[37]

Misunderstanding or mislabeling these materials often leads to improper disposal and **false consumer confidence**. Many end up in the same waste streams as conventional plastics, contributing to pollution under a different guise.[38]

Human Behavior and Unseen Plastic Spread

Some of the most pervasive pathways for microplastic contamination stem from **seemingly harmless actions** in daily life.

Brushing Teeth and Washing Hair

Toothbrushes and combs made from plastic, along with personal care products containing polymer ingredients, shed particles with each use. As these wash down drains, they enter wastewater systems where they are often too small to be captured by standard filters.[39]

Sweeping and Dusting

Dry sweeping can kick up settled microplastics into the air, increasing the chance of inhalation. Wet cleaning methods—mopping or wiping with damp cloths—are recommended to capture particles more effectively.[40]

Dish-washing and Kitchen Use

Using plastic utensils, cutting boards, and storage containers contributes to mechanical wear and tear. Microscopic shavings from chopping or microwaving food in plastic containers eventually enter food or rinse into drains.[41]

The Need for Global Monitoring Systems

Despite growing awareness, **comprehensive tracking of microplastics remains inadequate**. Most countries lack:

- Standardized testing protocols.
- Monitoring stations for air, soil, and water.

- Public health databases on exposure and effects.[42]

Some international efforts have emerged. The **UNEP (United Nations Environment Programme)** and **NOAA (National Oceanic and Atmospheric Administration)** support research and global cooperation. However, a centralized database and mandatory monitoring laws remain lacking.[43]

Citizen Science and Open Data

Grassroots efforts have filled some gaps. Apps and initiatives like:

- **Plastic Tracker.**
- **My Little Plastic Footprint.**
- **The Plastic Soup Foundation.**

...allow individuals to log plastic waste sightings, sample their environments, and share data. These efforts are valuable, but **governmental and industrial accountability** are essential for meaningful change.[44]

Final Thoughts: The Synthetic Saturation of Our World

Nano and microplastics are not just pollutants—they are now **integrated into the biosphere**. From the oceans to the soil, from food to air, and even within human bodies, these particles form a parallel ecology—synthetic, persistent, and globally distributed.

This omnipresence poses both scientific and ethical challenges:

- How do we manage what we can't see?
- Who is responsible for mitigating these "invisible invaders"?
- What actions will we take to restore balance?

Chapter 3 has shown that plastic contamination is not isolated or episodic—it is universal. Recognizing this truth is a critical step before action. In the next chapter, we turn from presence to **origin**, investigating where these plastics come from and how we might stem their flow at the source.

Chapter 4: Sources of Contamination

Introduction: From Creation to Contamination

Micro and nanoplastics don't appear out of nowhere. Their presence in ecosystems, food, and human tissue stems from identifiable, traceable sources that stem directly from our patterns of consumption and industrial design. Unlike some pollutants that are accidental byproducts, plastic particles are a predictable consequence of how we use, wear, wash, dispose of, and interact with synthetic materials. This chapter traces the full landscape of plastic particle contamination—beginning with deliberate production and ending with passive degradation—revealing the many sources that fuel the global plastic crisis.[1]

Primary vs. Secondary Microplastics

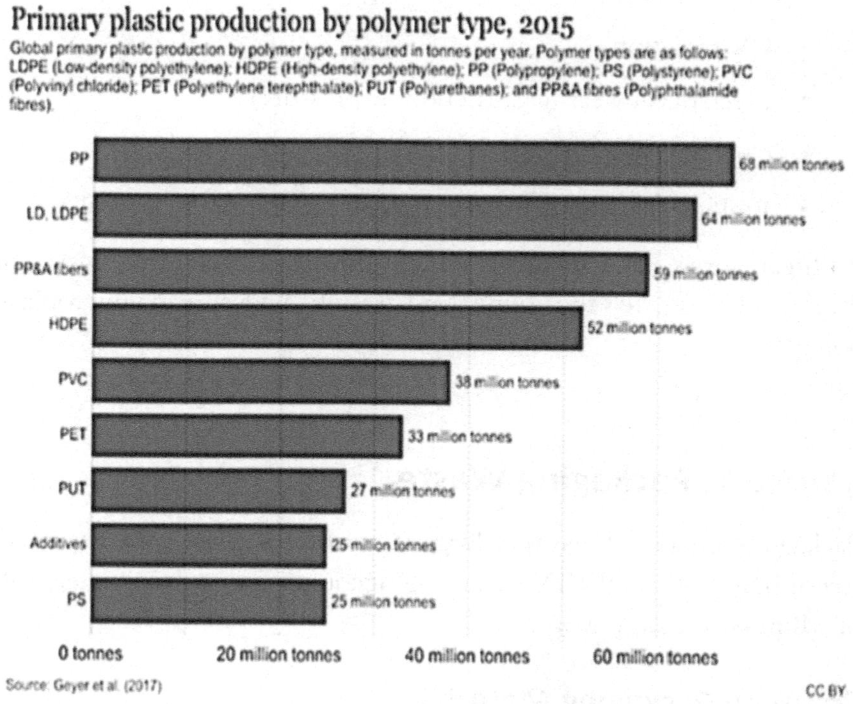

Figure 6: Primary Plasctic Production (Polymer Type)
Source: Our World In Data, CC BY 3.0 <https://creativecommons.org/licenses/by/3.0>, via Wikimedia Commons

Primary microplastics are intentionally manufactured small particles, created for specific industrial or consumer purposes. These include:

- Microbeads in exfoliants and toothpaste (now banned in many countries).
- Resin pellets (also known as nurdles) used in plastic manufacturing.

- Industrial abrasives used in sandblasting or ship cleaning.

Secondary microplastics, in contrast, are the byproduct of the breakdown of larger plastic items through:

- UV radiation.
- Mechanical wear (e.g., friction or pressure).
- Chemical degradation.
- Biological action (e.g., ingestion and excretion by animals).

Both categories enter the environment through different pathways, but they contribute to the same end result: widespread microplastic pollution.[2]

Source 1: Packaging Waste

Packaging remains the **single largest source** of plastic waste globally. According to the OECD, packaging accounts for approximately **40% of all plastic usage** worldwide.[3]

Common Packaging Materials

- Polyethylene terephthalate (PET): used in beverage bottles and food trays.
- High-density polyethylene (HDPE): used for milk jugs and detergent bottles.
- Polypropylene (PP): used for yogurt cups and bottle caps.

Mechanism of Particle Generation

- Improper disposal (littering, overflowing bins).

- Degradation in sunlight and weather.
- Mechanical friction in landfills or streets.

Plastic bags snag on trees, plastic wrap crumbles in gutters, and bottles shatter into shards—each step breaking items into smaller fragments.[4]

Extended Producer Responsibility (EPR)

Some countries and regions have begun to enforce EPR laws requiring companies to manage the lifecycle of their packaging. However, implementation is inconsistent and often lacks enforcement.[5]

Source 2: Synthetic Clothing and Fibers

Laundry has become a major contributor to the microplastic problem. When we wash clothes made of synthetic fibers—polyester, nylon, acrylic—**microfibers shed** from the fabric surface and escape into wastewater systems.

Key Stats

- One load of synthetic laundry can release up to **700,000 fibers.**
- Fleece jackets are among the worst offenders.
- Older garments shed more due to fiber fatigue.[6]

Wastewater Treatment Gaps

Although some microfibers are captured in wastewater treatment plants, many are not. Those that escape enter rivers, lakes, and oceans, while those that are trapped in sludge often **re-enter the environment** through agricultural fertilizer.[7]

Solutions

- Install washing machine filters.
- Use washing bags like Guppyfriend.
- Shift consumer demand toward natural fibers like hemp, cotton, and wool.

However, fast fashion trends and affordability of synthetic materials continue to drive microfiber pollution.[8]

Source 3: Tire Wear and Road Dust

Tires are made not only of rubber but also of **synthetic polymers** such as styrene-butadiene rubber. As tires roll over pavement, they **shed fine particles**—a mixture of rubber, plastic, metals, and road dust.

Estimates and Impact

- Tire wear accounts for **up to 30% of microplastic pollution in freshwater systems.**
- Urban runoff during rainstorms washes particles into storm drains and waterways.
- Airborne tire dust is increasingly linked to **respiratory health risks.**[9]

Vehicles as Vectors

It's not just tires—brake pads, paint coatings, and plastic vehicle parts also degrade. Collectively, vehicles represent one of the largest **non-point sources** of microplastics.[10]

Mitigation Technologies

- Roadside retention ponds.
- Street sweeping.
- Low-abrasion tires.
- Public transit and walking infrastructure to reduce car dependence.

Tire wear is a particularly insidious form of pollution because it occurs with **every mile driven**, without any visible trash or obvious "waste." [11]

Source 4: Industrial Processes and Manufacturing

Factories are significant contributors to microplastic pollution, even when plastic isn't their primary product. Manufacturing environments that use plastic resins, polymers, or plastic-coated components can shed particles during production, transport, or cleanup.[12]

Resin Pellet Loss (Nurdle Spills)

Plastic nurdles—small pellets melted to form products—are lost in large numbers during transport or handling. These spills occur:

- During transfer from ships to factories.
- In rail yards and loading docks.
- On factory floors, where spills often go uncleaned.

These pellets enter waterways through storm drains or are carried by wind. In some coastal areas, they represent one of the most common types of marine plastic pollution.[13]

Mechanical Abrasion

Machinery that cuts, sands, molds, or extrudes plastics creates microscopic shavings. These are rarely captured effectively and often enter wastewater or are swept into outdoor environments.[14]

Textile and Rubber Plants

Even non-plastic industries like textile or rubber manufacturing often use **polymer coatings, adhesives, or sealants**, which degrade and shed microplastics over time. This occurs in:

- Conveyor belts.
- Protective gear and gloves.
- Plasticized fabrics or foams.[15]

Worker Exposure

Industrial workers may inhale plastic dust or absorb compounds through their skin. Personal protective equipment (PPE) and air filtration can reduce exposure but are not universally applied.[16]

Source 5: Agricultural Inputs and Practices

While agriculture is typically viewed as a plastic victim (via soil contamination), it is also a source. From seed to harvest, modern farming uses plastic extensively.[17]

Plastic Mulch Films

To control weeds and retain moisture, many farms use thin plastic sheeting. This:

- Fragments under UV and mechanical stress.
- Leaves behind particles even when removed.
- Contributes to soil degradation.[18]

Greenhouse Covers and Irrigation Pipes

Over time, plastic greenhouse films and pipes weather and degrade, breaking down into microscopic fragments that accumulate in the soil.[19]

Fertilizer and Pesticide Carriers

Microplastics are sometimes used as **slow-release carriers** for fertilizers and pesticides, embedding chemical payloads within polymer matrices that degrade over time.[20]

Sewage Sludge (Biosolids)

Many farms apply treated sewage sludge to fields as fertilizer. This sludge often contains significant concentrations of microplastics, introducing them directly into the soil and crop root zones.[21]

Source 6: Personal Care and Cleaning Products

Despite increasing regulation, microplastics still appear in many everyday products.[22]

Cosmetics

Although microbeads have been banned in countries like the U.S., U.K., and Canada, other plastic-based ingredients remain:

- Polyethylene (PE).
- Acrylates Copolymers.
- Polyquaterniums.

These compounds act as:

- Film-formers in mascara.
- Thickeners in lotions.
- Texturizers in shampoos.[23]

Toothpaste and Exfoliants

Some brands continue to use **non-banned particles** or substitute with materials that degrade poorly. These microplastics bypass wastewater treatment and enter waterways.[24]

Household Cleaners

Floor polish, furniture sprays, and laundry detergents may contain polymer-based compounds for shine or durability. These often fragment over time or remain in wastewater.[25]

Hand Soaps and Sanitizers

Liquid soaps may use microplastic particles for texture or viscosity, and alcohol-based sanitizers with polymer thickeners break down into smaller residues.[26]

Source 7: Degradation of Waste in the Environment

Even when not in active use, plastics continue to pollute. Items discarded into the environment undergo natural and artificial degradation processes, generating a **constant stream of secondary microplastics**.[27]

Sunlight and UV Radiation

Sunlight causes photo-oxidation, which:

- Breaks polymer chains.
- Creates brittle, crack-prone plastic surfaces.
- Facilitates fragmentation into micro- and nanoplastics.[28]

Mechanical Stress

Wind, waves, and friction from sand or sediment further grind plastic debris into smaller pieces, especially along coastlines and riverbeds.[29]

Biological Activity

Insects, rodents, birds, and marine organisms often chew, digest, or mechanically damage plastic litter. Their activity contributes to accelerated fragmentation.[30]

Landfills and Dumpsites

Buried plastic undergoes anaerobic degradation, which is slower but still active. Landfill leachate—liquid runoff—has been found to contain high levels of plastic particles, which can:

- Contaminate groundwater.
- Overflow into local streams or sewers.[31]

Even so-called "safe disposal" sites contribute to long-term pollution, acting as time-release systems for microplastic contamination.[32]

Atmospheric Transport of Plastic Dust

In recent years, scientific attention has turned to **plastic in the air**. Atmospheric deposition of microplastics occurs through:

- Urban dust.
- Wind-borne fibers.
- Combustion byproducts.[33]

Indoor Air

Indoor spaces often have higher plastic particle concentrations than outdoor air due to:

- Synthetic carpets and upholstery.
- Electronics and cables.
- Clothing and blankets.[34]

Air Conditioning and Ventilation

HVAC systems can redistribute or concentrate microplastics in closed environments. Filters trap some particles but not all.[35]

Outdoor Air

Studies in urban, rural, and remote areas show airborne plastic deposition on:

- Soil.
- Lakes.
- Crop fields.
- Snow-packs and glaciers.[36]

The air we breathe is now a **plastic vector**, not just for local exposure but also for global distribution.[37]

Source 8: Marine Activities and Ocean-Based Inputs

A large share of oceanic plastic pollution does not originate from land —it's generated directly at sea. Marine-based activities contribute both large debris and microplastics to the aquatic environment.[38]

Fishing Gear and Lines

Discarded, lost, or degraded fishing equipment—such as:

- Nets made of nylon or polyethylene.
- Mono-filament fishing lines.
- Buoys and ropes.

These items shed fibers and fragments that accumulate on ocean floors or drift through marine ecosystems. Ghost nets also entangle marine life and degrade mechanically.[39]

Ship Paints and Coatings

Marine vessels use **plastic-based paints** with anti-fouling agents to prevent barnacle growth. These coatings release microplastics:

- Through abrasion against docks.
- When sandblasted for maintenance.
- During collisions or storms.[40]

Cleaning Agents and Bilge Water

Ships often use cleaning agents with synthetic polymers. When discharged—legally or illegally—they contribute microplastics to open waters.[41]

Ballast Water

Ballast water taken on in polluted harbors and released elsewhere transports microplastics (and invasive species) across the globe.[42]

Source 9: Tourism and Recreational Activities

Beaches, parks, hiking trails, and other tourist hotspots often suffer from acute plastic loading during high visitation periods.[43]

Beach Debris

Plastic cups, straws, wrappers, and bags degrade rapidly in UV-rich environments. These items fragment into microplastics that enter sand, water, and local fauna.[44]

Camping and Hiking Waste

Synthetic tents, raincoats, backpacks, and disposable utensils break down and leave behind residues. Fires may also incompletely burn plastics, releasing **toxic combustion particles**.[45]

Ski Resorts and Snow Toys

Recent studies have detected microplastics even in **remote alpine snow**, attributed partly to ski waxes, equipment, and clothing as well as atmospheric transport.[46]

Resort Waste Management

Some resorts lack proper waste collection infrastructure, especially in low-income or developing areas, leading to open dumping and local water contamination.[47]

Source 10: Mismanaged Recycling Systems

Ironically, recycling—often seen as a solution—can itself be a source of contamination if poorly managed.[48]

Mechanical Recycling

During shredding, grinding, or washing of plastics:

- Dust and micro-fragments are produced.
- Wastewater from the process contains microplastics.

- Residual contamination is not fully removed from recycled products.[49]

Plastic Contamination

Cross-contamination of plastic types can cause recycled materials to break down more easily or shed particles during use.[50]

Export of Waste

Many countries export plastic waste for processing elsewhere, often to regions with **lax environmental oversight**. This can lead to:

- Illegal dumping.
- Burning of non-recyclables.
- Runoff from open sorting yards.[51]

Downcycled Products

Recycled materials used in textiles, furniture, or construction may continue to shed microplastics, especially when mechanically stressed.[52]

Source 11: Fire Events and Plastic Combustion

Whether natural (wildfires) or artificial (landfill fires, house fires), burning plastic doesn't eliminate it—it transforms it.[53]

Combustion Residue

Plastics that burn incompletely can generate:

- Nano-sized soot particles.
- Polycyclic aromatic hydrocarbons (PAHs).
- Micro-ash laden with synthetic residues.[54]

These residues settle into soil, water, and buildings, especially in low-income or disaster-prone areas.[55]

Household Fires

Burning upholstery, electronics, and plastic pipes indoors creates localized hotspots of airborne plastic debris, which may persist in dust for years.[56]

Open Burning of Trash

In many parts of the world, household trash is burned in open pits. This includes plastic wrappers, packaging, and bags. This practice is especially common in rural or informal settlements and creates:

- Visible smoke with microplastic particles.
- Soil and water contamination nearby.
- Increased cancer and respiratory disease risks.[57]

Integrating the Web of Contamination

All these sources—packaging, clothing, tires, industry, agriculture, cosmetics, tourism, marine activities, and even recycling—interact to form a **complex and overlapping web** of contamination.[58]

- A synthetic fiber shed in one city may enter a wastewater treatment plant and be spread as sludge on a distant farm.
- Microplastics in air from a vehicle's tires can land on crops or infiltrate urban lungs.
- A plastic fragment degraded in a landfill may become nanoplastics that leach into aquifers and drinking water systems.

This inter-connectivity makes mitigation difficult—but not impossible.

The Invisible Spread: Nanoplastics

While microplastics get the spotlight due to their visibility and media attention, **nanoplastics** are arguably more dangerous and harder to trace. Nanoplastics are often:

- Formed from mechanical or UV degradation of microplastics.
- Generated during high-heat processes like burning or recycling.
- Produced by wear and tear in consumer electronics or medical devices.[59]

Because they are smaller than many cells, nanoplastics:

- Bypass biological barriers (e.g., gut lining, blood-brain barrier).
- Exhibit high surface reactivity.
- Bind with heavy metals and organic pollutants.
- Are extremely difficult to filter from air or water.[60]

The very features that make nanoplastics chemically interesting also make them biologically threatening.[61]

Final Thoughts

The sources of micro- and nanoplastic contamination are not abstract—they are directly tied to our daily lives, industries, and social systems. Each synthetic shirt, tire rotation, convenience wrapper, and cosmetic product plays a part in the broader narrative.

This chapter has laid bare the mechanisms of contamination. The next chapter will pivot toward solutions—showing how individuals can protect themselves and how societies can begin detoxifying the environment.

Chapter 5: Pathways into the Human Body

Introduction

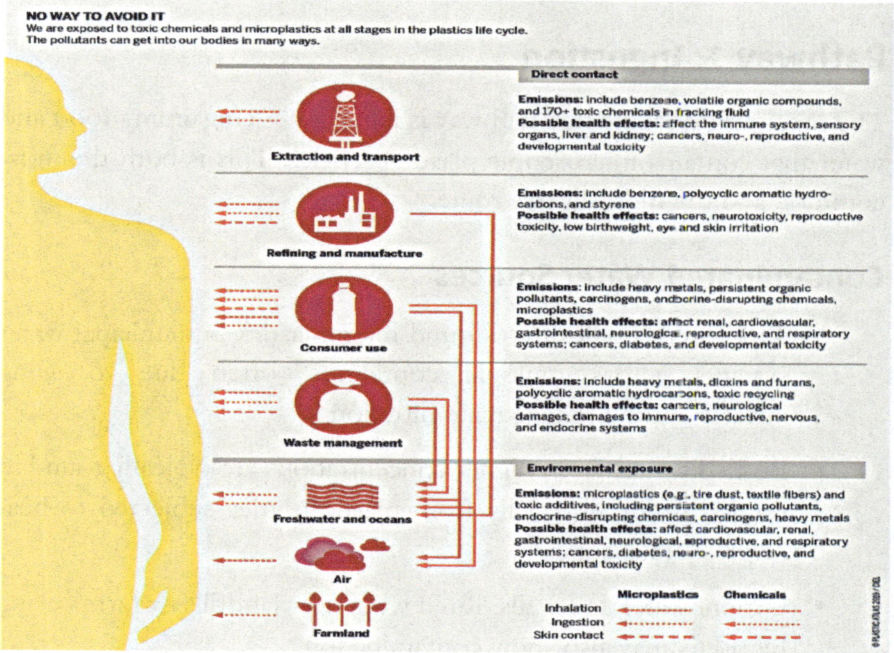

Figure 7: Exposure to Microplastics
Source: Graphic: PLASTIC ATLAS | Appenzeller/Hecher/Sack, CC BY 4.0 <https://creativecommons.org/licenses/by/4.0>, via Wikimedia Commons

Nano and microplastics are not just environmental pollutants—they are internal pollutants. As these particles become increasingly prevalent in our air, water, and food, researchers have begun mapping how they enter the human body and what happens next. This chapter focuses on the physical pathways by which plastic particles infiltrate our systems and identifies key exposure points, risk factors, and mechanisms of internal absorption. Understanding these entry routes is a critical first step in mitigating exposure and formulating effective public health responses.[1]

Pathway 1: Ingestion

The most well-documented pathway is ingestion—consuming food and water that contains microscopic plastic particles. This is both the most common and the most diverse route.

Contaminated Water Sources

- Tap water: Studies have found microplastics in municipal water systems across multiple continents, often due to aging infrastructure or inadequate filtration.[2]
- Bottled water: Even higher concentrations are typically found in bottled water, especially from plastic bottles subjected to heat or pressure changes.[3]
- Groundwater and wells: Rural wells near landfills or farms using biosolids may also show contamination.[4]

Food Contamination

Plastic particles have been detected in:

- Shellfish: Filter feeders like mussels and oysters absorb waterborne microplastics.[5]
- Fish: Species across the food chain contain microplastics in their tissues and organs.[6]

- Salt: Sea salt harvested from oceans often contains plastic residues.[7]
- Honey and beer: Contaminated processing equipment and water sources introduce microplastics even into processed products.[8]

Packaging Leachates

Plastic packaging breaks down during storage or cooking:

- Microwave meals: Heating plastic containers accelerates particle release.[9]
- Takeout food: Hot foods in styrofoam or thin plastic containers show higher contamination levels.[10]
- Packaged produce: Even non-processed foods can absorb microplastics from cling wrap or synthetic crates.[11]

Pathway 2: Inhalation

Airborne plastic particles are becoming an increasingly recognized vector of exposure. Inhalation may rival or even surpass ingestion in some urban and indoor environments.

Indoor Air

Sources of plastic in indoor air include:

- Synthetic carpets, furniture, and upholstery.
- Clothing made from polyester, nylon, and acrylic.
- Electronics, including printer emissions and cable sheathing.
- Household dust: A major carrier of microfibers and plastic fragments.[12]

Studies have shown that:

- We inhale tens of thousands of microplastic particles per year, mostly from indoor sources.[13]
- Indoor environments typically have higher microplastic concentrations than outdoors.[14]
- Infants may be more vulnerable due to floor-level exposure and frequent hand-to-mouth activity.[15]

Outdoor Air

While less concentrated than indoor air, outdoor environments still pose risks:

- Tire wear particles from traffic.
- Construction dust from insulation and sealants.
- Airborne plastic deposition from distant industrial sources.[16]

Urban residents and people living near highways, manufacturing zones, or incinerators may inhale significantly more plastic particles than rural dwellers.[17]

Pathway 3: Dermal Absorption

While the skin serves as a relatively strong barrier, some micro- and especially nanoplastics may still pass through or enter via wounds, pores, or hair follicles.

Personal Care Products

- Microbeads in facial scrubs and toothpastes can become embedded in skin tissue.[18]

- Lotions and cosmetics using polymer-based emulsifiers or preservatives may increase permeability.[19]
- Sunscreens with plastic-based dispersants or films may linger and be absorbed over time.[20]

Textile Contact

- Clothing made of synthetic fibers can shed particles that settle into skin or mucous membranes.
- Athletic wear, which is worn for long durations in sweaty conditions, may enhance transdermal exposure.[21]

Wound Entry

Nanoplastics are small enough to enter the bloodstream through broken or compromised skin, especially in environments where plastic dust or debris is airborne.[22]

Pathway 4: Medical and Pharmaceutical Exposure

Medical environments are one of the few places where nano- and microplastics are introduced deliberately, often for their biocompatibility and functional properties. However, unintended contamination also occurs.

Deliberate Uses

- Drug delivery systems: Nanoplastics are engineered to carry medication across biological barriers.[23]
- Sutures and implants: Plastic-based materials are used for their durability and resistance to degradation.[24]

- Medical tubing and devices: IV lines, catheters, and ventilator components often contain plasticized parts that may release particles.[25]

Unintended Exposure

- Disposable gloves and gowns shed particles during procedures.[26]
- Plastic packaging around sterile instruments may contaminate surfaces when opened.[27]
- Ventilated patients may be exposed to airborne microplastics if filters degrade or are improperly maintained.[28]

Biological Transport Mechanisms

Once nano- or microplastic particles enter the human body—via ingestion, inhalation, dermal contact, or medical exposure—they do not simply pass through passively. Instead, a number of **transport and uptake mechanisms** determine their fate within biological systems.

1. Gastrointestinal Uptake

Particles ingested through food or drink encounter several possible fates:

- **Excretion**: Many larger microplastics pass through the gastrointestinal (GI) tract and are expelled in feces.[29]
- **Mucosal Interaction**: Some particles may lodge in mucosal linings, triggering inflammation or immune responses.[30]

- **Translocation**: Nanoplastics, due to their small size, can cross the epithelial barrier of the intestines and enter systemic circulation via:
  - **M cells** in Peyer's patches.[31]
  - **Paracellular Transport** between epithelial cells.[32]
  - **Endocytosis** by enterocytes.[33]

2. Respiratory Absorption

Particles inhaled into the lungs may:

- **Settle in upper airways** and be expelled through mucociliary clearance.[34]
- **Penetrate Alveolar Sacs**, where nanoplastics may enter the bloodstream directly.[35]
- **Trigger inflammation**, leading to conditions such as asthma, bronchitis, or fibrosis.[36]

In laboratory studies, inhaled nanoplastics have been shown to accumulate in lung tissues and travel via the lymphatic system.[37]

3. Dermal Transport

While full skin penetration by microplastics is uncommon, **nanoplastics** may:

- Pass through sweat ducts or hair follicles.[38]
- Diffuse into dermal and subdermal layers.[39]
- Migrate to lymph nodes via macrophages.[40]

This pathway is particularly concerning for occupational exposure or repetitive use of topical products containing nano-compounds.[41]

4. Placental Transfer

Recent studies have confirmed the presence of micro- and nanoplastics in **human placenta.**[42] These particles likely translocate through maternal blood and accumulate in placental tissue, raising concerns about fetal exposure during critical stages of development.

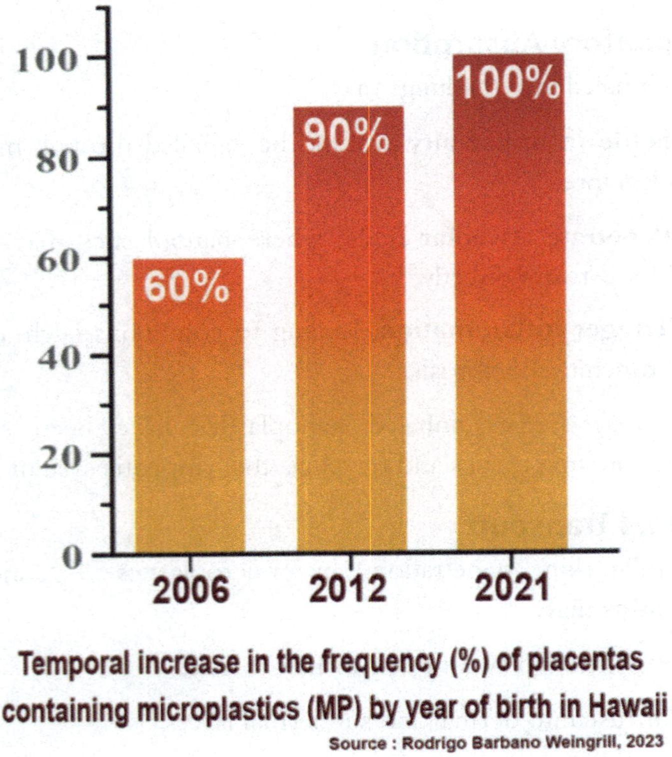

Figure 8: Placentas Containing Microplastics
Source: Lamiot, CC BY-SA 4.0 <https://creativecommons.org/licenses/by-sa/4.0>, via Wikimedia Commons

5. Blood-Brain Barrier Crossing

Nanoplastics may be capable of penetrating the **blood-brain barrier**, especially when coated with biofilms or chemical additives that mimic biological molecules.[43] This ability raises significant concerns regarding neurotoxicity and long-term cognitive effects.

Cellular Interactions and Uptake

Once in the bloodstream or tissues, plastic particles interact with various cellular components.

Phagocytosis by Immune Cells

Macrophages and dendritic cells may engulf nanoplastics, mistaking them for pathogens. This action can lead to:

- Inflammation.[44]
- Cytokine production.[45]
- Cell death (apoptosis or necrosis).[46]

Chronic immune activation due to persistent plastic exposure may weaken host defenses or lead to autoimmune disorders.[47]

Oxidative Stress

Plastic particles often generate **Reactive Oxygen Species (ROS)** when they interact with cellular membranes. This oxidative stress can:

- Damage DNA.[48]
- Accelerate aging of tissues.[49]
- Increase risk of cancer.[50]

Biofilm Formation

Nanoplastics may accumulate organic molecules or bacteria on their surfaces, forming **eco-coronas** or **biofilms**. These altered particles can evade immune detection or cause:

- Altered cell signaling.[51]
- Endothelial dysfunction.[52]
- Bacterial infections.[53]

Endocrine Interference

Additives in plastics (like BPA, phthalates, or flame retardants) can leach into cells and bind to hormone receptors, disrupting:

- Estrogen or androgen signaling.[54]
- Thyroid function.[55]
- Cortisol regulation.[56]

Dose and Duration: How Much Is Too Much?

The biological impact of plastic particles depends on:

- **Size**: Nanoplastics are more readily absorbed than microplastics.[57]
- **Shape**: Fibers and irregular particles may persist longer in tissues.[58]
- **Composition**: Different polymers and additives have unique toxicities.[59]
- **Dose**: Higher loads increase the probability of cellular interaction.[60]
- **Duration**: Chronic exposure, even at low levels, accumulates over time.[61]

Studies have found microplastic particles in human stool, blood, lungs, liver, kidney, and even brain tissue—suggesting that bioaccumulation is occurring silently across populations.[62]

Vulnerable Populations

Not all people experience plastic exposure equally. Certain groups face greater risks due to behavioral, physiological, or environmental factors.

Infants and Children

- Crawl and play near the ground, where dust and fibers accumulate.
- Chew on plastic toys or household items.
- Consume more water and food per unit of body weight than adults.

Studies have detected microplastics in **infant formula, baby bottles, and breast milk**, suggesting exposure begins from birth—or even before.[63]

Elderly

- Reduced detox capacity due to slower metabolism.
- Weakened immune defenses.
- Increased plastic exposure from medical devices or assisted living environments.[64]

Low-Income and Marginalized Communities

- Proximity to landfills, factories, or incinerators.
- Reliance on bottled water or fast food packaging.

- Limited access to air purification or filtered water.[65]

Industrial and Healthcare Workers

- Repeated contact with plastics in gloves, tubing, packaging, or equipment.
- High exposure to dust and fumes from manufacturing or sterilization processes.

Occupational exposure studies have revealed significantly higher microplastic burdens in workers from recycling facilities, textile plants, and hospitals.[66]

Long-Term Retention and Accumulation

Once nano- and microplastics gain entry into the human body and navigate cellular barriers, the question becomes: do they stay, and for how long?

Tissue Retention

Microplastic particles are often too large for efficient cellular clearance. Once deposited, they may:

- Remain trapped in organs such as the liver, spleen, lungs, or kidneys.[67]
- Accumulate in lymph nodes and vascular walls.[68]
- Embed within adipose (fat) tissue, shielding them from metabolic activity.[69]

Persistent Inflammation

Tissue retention of foreign bodies, including microplastics, commonly triggers chronic inflammation. This inflammatory environment:

- Damages healthy tissues.
- Disrupts normal repair processes.
- May lead to fibrosis, granuloma formation, or cellular mutations.[70]

Bioaccumulation

While traditional pollutants like mercury or PCBs have well-documented bioaccumulation curves, plastic particles exhibit similar behavior under certain conditions:

- Continuous exposure combined with poor clearance leads to gradual accumulation.[71]
- Re-ingestion of previously excreted microplastics through food or water can create feedback loops.[72]
- Additives or adsorbed chemicals on the plastic's surface (e.g., pesticides, heavy metals) may increase the total toxic burden.[73]

In short, the body lacks effective mechanisms to break down or eliminate plastic polymers—especially those in nano-form.[74]

Synergistic Toxicity: Plastics and Other Pollutants

Plastics rarely travel alone. In environmental and biological settings, they often interact with other harmful substances—making their combined effects more dangerous than each component alone.

Chemical Additives

Most commercial plastics contain added chemicals:

- **Plasticizers** like phthalates to increase flexibility.
- **Stabilizers** like lead compounds or organotins.
- **Colorants**, flame retardants, and antimicrobial agents.

When ingested or absorbed, these compounds may leach into tissue even if the plastic particle itself remains inert.[75]

Environmental Pollutants

Plastic surfaces attract and bind:

- Persistent organic pollutants (POPs).
- Polycyclic aromatic hydrocarbons (PAHs).
- Heavy metals like cadmium, lead, or arsenic.

This binding creates toxic "cocktails" that:

- Increase the ability of pollutants to cross cellular membranes.
- Act as endocrine disruptors, neurotoxins, or carcinogens.
- Complicate efforts to isolate and study individual exposure pathways.[76]

Biological Hitchhikers

Microplastics in the body may also carry:

- Pathogenic bacteria or viruses from contaminated water or surfaces.
- Antibiotic-resistant genes and microbes from wastewater environments.

This risk is particularly high for immunocompromised individuals, where colonization can lead to opportunistic infections.[77]

Scientific Frontiers: Tracking Plastics Inside the Body

Given the novelty and scale of plastic contamination, scientific tools and models are rapidly evolving to track how plastics move through the body.

Biomonitoring

Recent advances include:

- Blood and urine tests for plastic-associated chemicals.
- Fecal microplastic quantification.
- Biopsies of placental and organ tissues to assess plastic accumulation.[78]

Animal Models

Rodent and fish models show:

- Liver enlargement and fatty deposits after plastic exposure.
- Cognitive impairment in offspring when mothers are exposed prenatally.
- Hormonal imbalances and reproductive toxicity.[79]

In Vitro Studies

Cell culture experiments have revealed:

- Cell membrane disruption and death from nano-sized particles.

- DNA fragmentation and mutagenesis.
- Altered gene expression related to detox pathways.

These studies, while not yet fully translatable to humans, offer strong early warnings.[80]

Emerging Technologies

New tools to study plastic pathways include:

- Fluorescent tagging of nanoplastics for in vivo tracing.
- Magnetic microplastics to observe systemic circulation.
- Artificial organs (organoids) to simulate barrier penetration.[81]

Public Health Implications

The confirmation of plastic pathways into the human body has profound implications for healthcare, regulation, and environmental policy.

1. Risk Communication

Governments and health authorities must:

- Inform the public of major exposure sources.
- Create labeling systems for products free from microplastics.
- Educate communities on high-risk environments (e.g., indoor air, food packaging).[82]

2. Clinical Practice

Healthcare professionals should be aware of plastic-related symptoms, especially for:

- Autoimmune disorders.
- Hormonal imbalances.

- Chronic respiratory or gastrointestinal issues.

Integrative medicine may include screening for environmental toxin burden and offering lifestyle-based detox protocols.[83]

3. Regulation and Standards

Policymakers should enforce stricter controls on:

- Additives used in food packaging and personal care products.
- Emission levels from industrial and agricultural plastic sources.
- Wastewater filtration standards to remove microplastics before reintroduction into public systems.[84]

4. Vulnerable Groups

Targeted protections must be put in place for:

- Pregnant women and infants.
- Industrial workers.
- People living in environmental justice communities.[85]

Final Thoughts

The routes by which plastic enters the body are varied, subtle, and disturbingly commonplace. Ingestion, inhalation, dermal contact, and even medical exposure each play a role in the slow infiltration of these synthetic pollutants into our most vital biological systems.

More troubling still is what happens after entry: accumulation, inflammation, endocrine disruption, and synergy with other toxins. The body, evolved over millennia to handle organic contaminants, is not equipped to deal with polymers that mimic—but do not behave like—natural substances.

As scientific research catches up to the reality of exposure, a paradigm shift is needed. Just as we once learned the dangers of lead, asbestos, and tobacco, the pathways of micro- and nanoplastics demand urgent attention. The health of current and future generations depends on recognizing these threats and responding with bold, informed action.[86]

Chapter 6: Impact on Human Health

Introduction

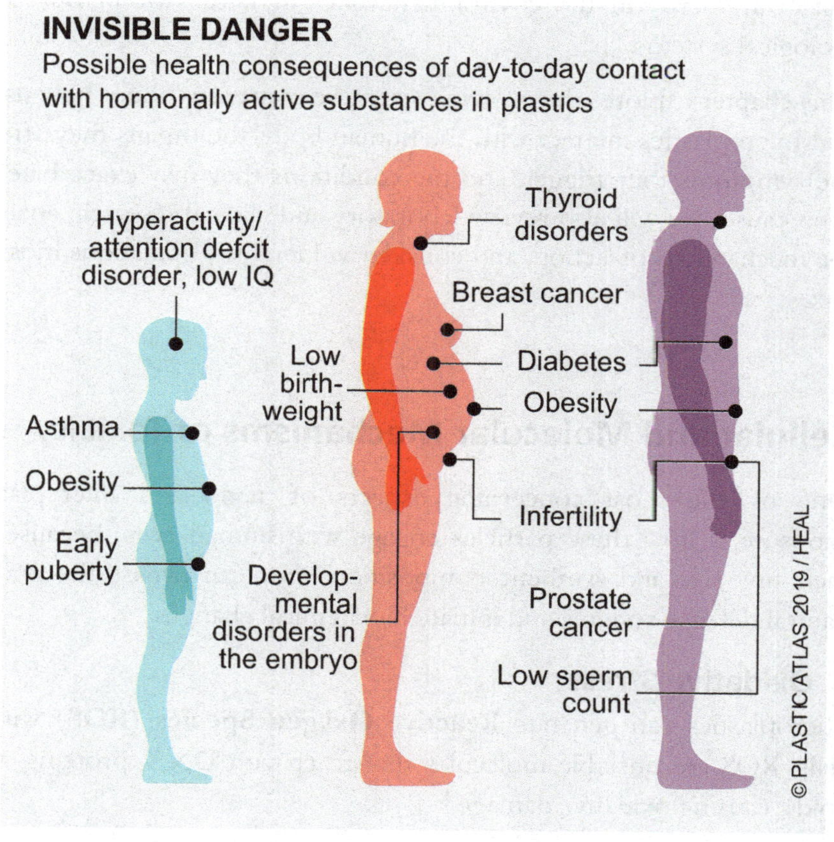

Figure 9: Health Consequences of Microplastics
Source: PLASTIC ATLAS | Appenzeller/Hecher/Sack, CC BY 4.0 <https://creativecommons.org/licenses/by/4.0>, via Wikimedia Commons

Nano and microplastics are no longer a distant environmental concern; they are increasingly recognized as internal pollutants with direct implications for human health. These synthetic fragments, often invisible to the naked eye, have been found in blood, lungs, placenta, breast milk, and even fecal samples—evidence that plastic is no longer just "out there" in the environment but "in here" within our own biological systems.

This chapter explores the evolving scientific consensus about how nano and microplastics interact with the human body, the organs they affect, the symptoms they trigger, and the conditions they may exacerbate or even cause. We will also review laboratory and clinical research, analyze the mechanisms of action, and consider vulnerable populations most at risk.

Cellular and Molecular Mechanisms of Toxicity

One of the most concerning aspects of nano and microplastic exposure is how these particles engage with human cells. Because of their tiny size and synthetic composition, they can bypass the body's natural defense systems and initiate biochemical changes.

1. Oxidative Stress

Microplastics can generate **Reactive Oxygen Species (ROS)** within cells. ROS are unstable molecules that react with DNA, proteins, and lipids, causing oxidative damage.[1]

- This damage is linked to aging, cancer, cardiovascular disease, and neurodegeneration.
- Chronic oxidative stress also weakens immune response and promotes chronic inflammation.[2]

2. Inflammation

When the body recognizes plastic particles as foreign invaders, immune cells—particularly macrophages—attempt to engulf them via **phagocytosis**.[3]

- This process often fails, especially for larger or irregular particles, leading to prolonged immune activation.
- Persistent inflammation is a root cause of numerous diseases, including diabetes, arthritis, and inflammatory bowel disease.[4]

3. Endocrine Disruption

Many plastics contain **additives** like bisphenol A (BPA), phthalates, and flame retardants.[5]

- These substances mimic natural hormones and bind to hormone receptors, disrupting endocrine signaling.
- Conditions associated with endocrine disruption include infertility, thyroid disorders, obesity, and early puberty.[6]

4. Cellular Apoptosis and Necrosis

In vitro studies have shown that exposure to nanoplastics can lead to:

- Cell membrane rupture.
- Mitochondrial dysfunction.
- Programmed cell death (apoptosis).[7]

5. Genotoxicity

Some plastic particles, particularly those carrying heavy metals or chemical pollutants, have been shown to damage DNA, potentially leading to mutations and cancer.[8]

Organ-Specific Impacts

Different organs are affected depending on the route of exposure and the characteristics of the plastic particles (e.g., size, shape, composition).

Lungs

- Inhaled microplastics can settle in the alveoli, impair gas exchange, and trigger asthma-like symptoms.[9]
- Occupational exposure has been linked to lung fibrosis and granuloma formation.[10]
- Studies show elevated inflammatory markers in lung tissue exposed to synthetic fibers.[11]

Liver

- As a central organ for detoxification, the liver accumulates plastic particles absorbed via the gut.
- Animal studies show liver enlargement, fat deposits, and altered enzyme activity.[12]
- Plastic exposure may increase risk of non-alcoholic fatty liver disease (NAFLD).[13]

Kidneys

- Kidneys filter blood and excrete waste, making them vulnerable to waterborne nanoplastics.
- Evidence suggests impaired renal filtration and increased oxidative stress markers.[14]

- Chronic exposure may elevate the risk of kidney disease and electrolyte imbalances.[15]

Gastrointestinal Tract

- Particles can disrupt the mucosal barrier and microbiome composition.
- Dysbiosis (imbalance of gut bacteria) has been linked to systemic inflammation, obesity, and mental health issues.[16]
- Rodent studies show decreased nutrient absorption and increased gut permeability ("leaky gut").[17]

Brain and Nervous System

- Emerging research indicates that nanoplastics may cross the **blood-brain barrier.**[18]
- This raises concerns about neuroinflammation, memory impairment, and neurodevelopmental delays.
- Animal studies show behavioral changes and neurotransmitter imbalances following plastic exposure.[19]

Reproductive and Developmental Concerns

One of the most alarming dimensions of plastic toxicity involves reproductive health and fetal development.

Hormonal Interference

Plastic-associated endocrine disruptors have been shown to:

- Reduce sperm count and motility.[20]

- Alter menstrual cycles and ovulation patterns.[21]
- Increase risk of polycystic ovary syndrome (PCOS) and endometriosis.[22]

Fetal Exposure

Microplastics have been found in the **placenta**, suggesting prenatal exposure.[23]

- Disruption of hormonal signaling during pregnancy can affect organogenesis and fetal growth.
- Animal studies show lower birth weights, altered sex hormone levels, and developmental delays in offspring.[24]

Transgenerational Effects

There is growing concern that plastic exposure may induce epigenetic changes—heritable modifications in gene expression that don't involve changes to DNA sequence.

- Such changes can affect not only the exposed individual but also their children and grandchildren.[25]

Immunological Disruption

Micro- and nanoplastics interact with the immune system in complex ways. These synthetic intruders are often misidentified as biological threats, leading to immune activation that may become chronic.

Innate Immune Response

- Macrophages and neutrophils—the body's first line of defense —attempt to phagocytose plastic particles.

- This interaction is inefficient, especially with fibrous or irregular particles, leading to "frustrated phagocytosis," a process that releases enzymes and inflammatory mediators into surrounding tissues.[26]
- The result: local tissue damage and signaling cascades that provoke systemic inflammation.

Adaptive Immunity

- Prolonged exposure to plastic-related antigens may dysregulate the adaptive immune system.[27]
- T-cells may become overactive, misfire, or exhaust, impairing their ability to respond to infections.
- B-cells may produce autoantibodies, potentially contributing to autoimmune diseases.[28]

Cytokine Storm Potential

- Plastic particles have been shown in lab studies to elevate cytokine production—especially TNF-α, IL-6, and IL-1β.[29]
- These inflammatory messengers, when produced in excess, can contribute to tissue degradation and systemic inflammatory responses.

Allergies and Hypersensitivities

- Plastics may contribute to the rising prevalence of allergic diseases.[30]
- Microplastics embedded in the nasal cavity or lungs may prime the immune system for hyperreactivity.

- Anecdotal clinical data points to increased dermatitis in individuals exposed to synthetic fibers and polymers.[31]

Metabolic Effects and Obesity

Plastic-related chemicals such as BPA and phthalates are classified as **obesogens**—compounds that alter lipid metabolism and promote fat storage.

Hormonal Disruption

- BPA mimics estrogen, a hormone deeply involved in fat regulation and energy metabolism.[32]
- Phthalates reduce testosterone and interfere with thyroid hormones, both of which are critical to maintaining a healthy weight and metabolism.[33]

Adipogenesis Activation

- Studies show that plastic chemicals increase the expression of genes that drive adipogenesis—the creation of new fat cells.[34]
- This may lead to an increase in visceral fat, which is associated with higher risks of heart disease and diabetes.

Insulin Resistance

- Plastic-related pollutants have been linked to impaired insulin signaling.[35]
- This contributes to **metabolic syndrome**, a cluster of conditions including hypertension, hyperglycemia, and dyslipidemia.

Microbiome and Weight Gain

- The gut microbiome plays a crucial role in calorie extraction and fat storage.
- Microplastic-induced dysbiosis may promote microbial species associated with weight gain and inflammation.[36]

Cardiovascular Health

The heart and vascular system are indirectly affected by plastic particles through inflammation, oxidative stress, and hormonal disruption.

Blood Pressure Regulation

- Endocrine-disrupting chemicals can affect the renin-angiotensin system, which regulates blood pressure.[37]
- Persistent low-grade inflammation from plastic exposure may lead to vascular stiffening and increased systolic pressure.[38]

Endothelial Dysfunction

- Nanoplastics have been shown to cause endothelial cell damage in lab cultures.[39]
- Damaged endothelium is a precursor to atherosclerosis and thrombosis.

Circulating Plastic Particles

- In 2022, a groundbreaking study confirmed the presence of microplastics in human blood.[40]

- Plastic particles may travel through the bloodstream and deposit in narrow capillaries, leading to microvascular damage.

Heart Disease Risk

- The systemic inflammation triggered by microplastic accumulation is a recognized risk factor for cardiovascular events.[41]
- Additionally, indirect effects—such as metabolic syndrome and hormonal imbalance—compound this risk.

Neurological Effects

Research is increasingly suggesting that plastic exposure may affect brain health, cognition, and mental well-being.

Crossing the Blood-Brain Barrier

- Nanoplastics, due to their small size and surface properties, may penetrate the **blood-brain barrier (BBB)**.[42]
- This raises the possibility of direct neural exposure to synthetic contaminants.

Neuroinflammation

- Once inside the brain, plastic particles may trigger an immune response involving microglia, the brain's resident macrophages.[43]

- Chronic microglial activation has been linked to neurodegenerative conditions such as Alzheimer's and Parkinson's diseases.

Cognitive and Behavioral Changes

- Rodent studies show reduced memory performance and anxiety-like behavior following plastic ingestion.[44]
- Neurotransmitter levels—especially dopamine and serotonin—are altered in plastic-exposed animal models.[45]

Child Development

- Fetal and early childhood exposure to endocrine disruptors like BPA has been associated with:
    - Lower IQ scores.[46]
    - Attention-deficit/hyperactivity disorder (ADHD).[47]
    - Autism spectrum characteristics (correlational, not causal).[48]

Sleep and Mood

- Hormonal imbalances resulting from plastic exposure may impact sleep quality, mood regulation, and circadian rhythm.[49]
- Elevated cortisol levels and altered melatonin metabolism are under investigation in this context.

Cancer Risk and Plastic Exposure

While long-term epidemiological studies are still developing, there is increasing concern over the potential **carcinogenicity** of plastic exposure—particularly at the nano scale.

1. Chemical Additives as Carcinogens

- Certain plasticizers and stabilizers, such as **phthalates, vinyl chloride**, and **styrene**, are classified as *probable human carcinogens* by the International Agency for Research on Cancer (IARC).[50]
- Chronic exposure, especially through food and beverage packaging, increases the cumulative load of these compounds in the body.

2. Particle-Induced Carcinogenesis

- Nanoplastics that reach cellular DNA may induce genotoxic effects.[51]
- Studies on rodents have shown pre-cancerous lesions in the liver and colon after prolonged plastic ingestion.[52]
- Plastic particles coated with environmental toxins (e.g., heavy metals, dioxins) pose an even greater risk.[53]

3. Tumor Microenvironment

- Plastics can affect the tumor microenvironment by altering immune surveillance and supporting angiogenesis (new blood vessel formation).[54]
- This may enable tumor cells to evade detection and metastasize.

4. Hormone-Driven Cancers

- Endocrine disruptors have been implicated in hormone-sensitive cancers such as breast, prostate, and testicular cancer.[55]
- BPA, for example, has estrogenic effects that may promote the growth of estrogen receptor-positive tumors.[56]

Vulnerable Populations

The health impacts of nano and microplastics are not uniformly distributed. Certain populations are more susceptible due to physiological, environmental, or socio-economic factors.

Infants and Children

- Children have higher respiratory rates and lower body mass, leading to proportionally higher exposure.[57]
- Crawling on carpets and mouthing plastic toys increases ingestion and inhalation risks.
- Early-life exposure during critical development windows has disproportionate effects.

Pregnant Women

- The presence of microplastics in the **placenta** and **amniotic fluid** raises concerns about fetal health.[58]
- Hormonal fluctuations during pregnancy make women more susceptible to endocrine disruption.

Elderly Individuals

- Aging populations often have reduced liver and kidney function, impairing detoxification.[59]
- Pre-existing inflammatory conditions may be exacerbated by plastic exposure.

Workers in High-Exposure Occupations

- Textile workers, plastic manufacturers, and waste recyclers inhale high levels of microfibers.[60]
- Chronic exposure without adequate respiratory protection increases risk for pulmonary and systemic diseases.

Low-Income Communities

- Often located near industrial zones or waste sites with elevated plastic pollution.[61]
- May rely more on plastic-packaged foods, bottled water, and synthetic clothing.
- Have reduced access to healthcare and information regarding environmental risks.[62]

Synergistic Effects with Other Pollutants

Plastic particles rarely act alone. They often serve as vectors for other environmental toxins, compounding their health effects.

Heavy Metals

- Microplastics adsorb mercury, lead, cadmium, and arsenic from their surrounding environments.[63]
- When ingested, these metals can dislodge and enter the bloodstream, affecting the brain, kidneys, and bones.[64]

Persistent Organic Pollutants (POPs)

- Dioxins, PCBs, and PAHs can hitchhike on plastic surfaces.[65]
- These compounds are known to be immunotoxic, neurotoxic, and carcinogenic.
- Combined exposure accelerates systemic toxicity even at lower doses.[66]

Antibiotic Resistance Genes

- Microplastics in wastewater have been found to carry **antibiotic-resistant bacteria and resistance genes**.[67]
- This may contribute to the growing global health crisis of antimicrobial resistance (AMR).[68]

Gaps in Research and Regulatory Implications

The health risks of nano and microplastics are still being mapped. While the evidence is mounting, major gaps remain.

Lack of Standardized Detection

- There is no universally accepted method for detecting plastic particles in human tissues or fluids.[69]

- Different studies use varying definitions, sizes, and detection techniques, complicating comparisons.

Unknown Dose-Response Relationships

- How much exposure is too much? The threshold levels for harm remain unclear.
- Nanoplastics may be harmful at far lower doses than currently assumed.[70]

Interindividual Variation

- Genetic differences affect how people metabolize and respond to plastic toxins.
- Epigenetic factors and pre-existing conditions complicate risk assessment.[71]

Regulatory Inaction

- Most countries lack specific regulations for microplastics in food, air, or water.
- Industry lobbying and scientific uncertainty slow down legislative efforts.[72]

Future Directions for Health Protection

While full-scale solutions require systemic change, individuals and health institutions can take proactive steps.

1. Integrating Plastic Exposure into Clinical Histories

- Healthcare providers should ask about occupational exposure, diet, and plastic use patterns during check-ups.[73]
- This will help identify correlations between symptoms and environmental load.

2. Public Health Campaigns

- Governments can launch campaigns to educate the public about low-plastic living.
- Similar to tobacco warnings, packaging could include plastic toxicity information.[74]

3. Biomedical Innovation

- Researchers are developing nanosponges, detoxifying enzymes, and plastic-degrading microbes that may be used internally in the future.[75]

4. Longitudinal Studies

- We need cohort studies tracking plastic exposure and health outcomes over decades.
- This will establish stronger causal links and help prioritize regulatory actions.[76]

5. Environmental Justice Advocacy

- Protecting vulnerable communities must become a central goal.
- Disparities in exposure and health outcomes require targeted interventions.[77]

Final Thoughts

The story of plastic is no longer confined to the ocean or landfill—it is unfolding inside our bodies. From hormone disruption to neurological effects, micro and nanoplastics interact with human physiology in deeply complex and still unfolding ways. They are stealth invaders, entering silently but with consequences that could ripple across generations.

Understanding the impact of these particles is not just a scientific imperative but a moral one. Protecting human health in the era of synthetic pollution requires a multipronged approach—research, regulation, public education, and most importantly, individual and collective action.

As new evidence continues to emerge, one thing remains clear: plastics are not biologically inert. They are not passive litter. They are dynamic, reactive, and potentially dangerous. Recognizing this truth is the first step toward mitigating their harm and reshaping our future.

Chapter 7: Ecological and Biodiversity Effects

Introduction

Figure 10: Global Mismanaged Plastic Waste (Regional)
Source: Our World In Data, CC BY 3.0 <https://creativecommons.org/licenses/by/3.0>, via Wikimedia Commons

Nano and microplastic pollution presents one of the most insidious threats to ecosystems worldwide. Unlike larger debris that can be removed or observed, these particles are small enough to infiltrate biological systems from the ground up. The microscopic scale of nano- and microplastics allows them to bypass conventional filtration mechanisms, become embedded in sediments, ingested by wildlife, and enter food chains with far-reaching consequences. This segment examines the effects of these pollutants on ecological systems, starting with soil environments and progressing to freshwater habitats.

Microplastics in Soil Ecosystems

Soil ecosystems are often overlooked in discussions of plastic pollution, yet they are among the most affected. Plastics enter terrestrial environments via agricultural activities, plastic mulching, sewage sludge application, atmospheric deposition, and litter degradation. Once present, these particles interact with the complex web of microbial life and fauna that sustain terrestrial biodiversity and nutrient cycling.[1]

Disruption of Microbial Balance

The microscopic communities in soil—comprising bacteria, fungi, archaea, and protozoa—are fundamental to soil fertility.[2] Studies show that microplastics can disrupt these microbial populations in several ways:

- **Community shifts**: Certain bacteria, particularly those capable of colonizing plastic surfaces (plastisphere communities), tend to outcompete native soil microbes, leading to reduced diversity.
- **Reduced enzymatic activity**: Soil enzyme activities related to nitrogen fixation and organic matter decomposition are often diminished in plastic-contaminated soils.
- **Biofilm formation**: Microplastics provide new surfaces for microbial colonization, which may alter normal microbial interactions and biochemical pathways.

Such changes are not benign. Altered microbial function can decrease soil fertility, reduce organic matter decomposition, and impact plant health and agricultural productivity.[3]

Impacts on Soil Fauna

Earthworms, nematodes, collembolans, and mites form an integral part of the soil food web. Earthworms, in particular, play a key role in aeration, organic matter decomposition, and nutrient redistribution. Unfortunately, these organisms often ingest microplastics unintentionally while feeding.

- **Digestive obstruction**: Microplastic ingestion can physically block digestive tracts, reducing nutrient absorption and leading to starvation.
- **Reproductive issues**: Earthworms exposed to microplastics exhibit lower reproduction rates and delayed maturation.
- **Burrowing behavior**: Changes in soil structure and increased toxicity impair the natural burrowing behavior of these animals, which in turn affects soil turnover and aeration.

These disruptions cascade through the food web. Predators of soil fauna (e.g., birds and small mammals) are indirectly affected as prey availability and quality diminish.[4]

Effects in Freshwater Environments

Freshwater ecosystems serve as both passageways and sinks for nano- and microplastics. These pollutants reach lakes, rivers, and streams through urban runoff, wastewater discharge, and atmospheric deposition. Once introduced, plastics distribute quickly across water columns and sediment layers.[5]

Plankton and Primary Producers

Plankton—including phytoplankton and zooplankton—are the foundational producers and consumers in freshwater food webs. Their interactions with plastics reveal early-stage ecological risks.

- **Phytoplankton**: Microplastics reduce light penetration, hampering photosynthesis. Surface-bound contaminants also leach into the water, introducing toxic substances.
- **Zooplankton**: These tiny animals readily ingest microplastics due to their size and resemblance to natural food particles. Ingested plastics block nutrient absorption and affect growth and reproduction.

Since plankton are a primary food source for fish larvae and other aquatic species, any reduction in their population or health compromises the entire ecosystem structure.[6]

Trophic Transfer and Bioaccumulation

Small fish and macroinvertebrates that consume contaminated plankton become secondary vectors for microplastic movement up the food chain. Over time, plastic particles and their adsorbed toxins accumulate in tissues, leading to:

- **Toxicological effects**: Chronic inflammation, oxidative stress, and immune dysfunction.
- **Reproductive decline**: Altered hormone levels and reduced fecundity.

- **Behavioral changes**: Reduced predatory responses, slower swimming, and altered schooling behavior.

The transfer of these effects through multiple trophic levels destabilizes ecosystem resilience and increases extinction risk for sensitive species.

Impacts on Marine Ecosystems

Marine environments have become the most iconic symbol of plastic pollution. Images of sea turtles entangled in bags and birds with bellies full of bottle caps have captured public attention, but nano and microplastics pose a more hidden—and possibly more dangerous—threat. These particles are found in every layer of the ocean, from surface waters to deep-sea sediments, and across every marine habitat, including coral reefs, estuaries, and polar seas.[7]

Filter Feeders and Suspension Feeders

Organisms that feed by filtering water, such as mussels, oysters, baleen whales, and various planktonic creatures, are especially vulnerable.[8]

- **Ingestion and retention**: Filter feeders indiscriminately process large volumes of water and ingest suspended particles, including microplastics. Many species cannot effectively expel these particles.
- **Physical blockages and inflammation**: Ingested plastics cause internal abrasions and blockages, leading to reduced feeding efficiency and inflammation of tissues.
- **Nutritional dilution**: With stomach space occupied by indigestible plastic, animals feel "full" without receiving nutrients, leading to malnourishment over time.

These impacts are not just individual—filter feeders play a vital role in nutrient cycling, water clarity, and ecosystem engineering. Their decline affects entire coastal and benthic systems.[9]

Coral Reefs

Coral reefs are biodiversity hotspots and essential coastal protectors, but they are increasingly compromised by plastic pollution.[10]

- **Settling and smothering**: Microplastics settle onto coral surfaces and physically obstruct light and gas exchange.
- **Disease vectoring**: Plastics introduce pathogenic microbes, contributing to coral bleaching and necrosis.
- **Reproductive interference**: Plastics impair coral larvae development and settlement, threatening reef regeneration.

Because corals build the very foundation of reef ecosystems, their degradation leads to habitat loss for hundreds of species, including commercially important fish.[11]

Predatory Fish and Apex Species

Microplastics have been found in the digestive tracts, gills, and muscle tissues of large fish, sharks, and marine mammals.[12]

- **Gastrointestinal damage**: Abrasions, ulcers, and intestinal blockages lead to pain, reduced feeding, and organ failure.
- **Chemical contamination**: Plastics leach toxins like bisphenol A, PCB, and heavy metals into the host, disrupting endocrine and neurological systems.

- **Behavioral anomalies**: Disorientation, aggression, or apathy may result from neurotoxic effects, compromising survival and reproduction.

Since apex predators regulate lower trophic levels, their decline can lead to trophic cascades, where prey populations explode and ecosystems become unbalanced.[13]

Birds and Terrestrial Wildlife

The effects of nano and microplastics are not limited to aquatic ecosystems. Birds, mammals, and even insects encounter and ingest these particles through contaminated food, water, and nesting materials.[14]

Seabirds

Seabirds are particularly well-studied indicators of plastic pollution due to their widespread distribution and high plastic ingestion rates.[15]

- **Stomach Impaction**: Chicks fed plastic by parents suffer from gut impaction, stunted growth, and starvation.
- **Reduced Reproductive Success**: Plastic-laden nests and reduced adult foraging efficiency result in fewer fledged offspring.
- **Internal Damage and Contamination**: Plastics cause internal wounds and expose birds to endocrine-disrupting chemicals.

Hundreds of seabird species have documented microplastic ingestion, with many showing population-level effects.[16]

Terrestrial Mammals and Insects

Emerging research shows that plastics are reaching terrestrial wildlife through indirect pathways:[17]

- **Herbivores** consume plastic-contaminated vegetation or drink from polluted sources.
- **Predators** ingest plastic through prey that have bioaccumulated particles.
- **Scavengers and opportunists** (e.g., foxes, raccoons) forage in landfills and urban environments where plastic is abundant.
- **Insects**, such as bees, may incorporate microplastics into their nests or collect it on their bodies during foraging.

While more studies are needed, early findings suggest these exposures impair digestion, alter behavior, and reduce reproductive health in many species.[18]

Impacts on Amphibians and Reptiles

Amphibians and reptiles occupy unique ecological niches and are often sensitive bioindicators of environmental stress due to their permeable skin, complex life cycles, and dependence on both aquatic and terrestrial environments. The presence of nano and microplastics in their habitats can severely disrupt physiological functions and ecosystem dynamics.[19]

Amphibians

Amphibians such as frogs, toads, salamanders, and newts are particularly susceptible to contaminants during their larval (aquatic) stages. Plastics can affect them through multiple exposure routes:

- **Direct Ingestion**: Tadpoles ingest microplastic particles present in the water column. These can accumulate in the gastrointestinal tract, causing digestive blockage and nutrient absorption interference.[20]
- **Toxicity from additives**: Many plastics carry chemical additives (e.g., phthalates, bisphenol A) that can leach into amphibian tissues, disrupting hormone function.[21]
- **Developmental Abnormalities**: Chronic exposure to microplastics has been linked to limb deformities, altered metamorphosis timing, and immune suppression.[22]
- **Habitat Alteration**: Plastics in breeding ponds or wetland habitats may alter sediment composition and water quality, decreasing egg viability and larval survival.[23]

Since amphibians often serve as both predator and prey in ecosystems, their decline leads to cascading effects on insect populations, aquatic vegetation, and the species that feed on them, such as birds and snakes.[24]

Reptiles

Reptilian species, including turtles, snakes, and lizards, face growing threats from plastic contamination.

- **Turtles**: Marine turtles are well-documented victims of plastic ingestion, but freshwater and terrestrial turtles are also affected.

Ingested plastics mimic jellyfish or vegetation, leading to digestive obstruction, starvation, and even buoyancy issues, which prevent normal diving behavior.[25]

- **Lizards and Snakes**: These animals may consume prey that has ingested plastic, leading to secondary exposure. Although less studied, evidence suggests that endocrine disruption and liver damage may result from accumulated microplastics and associated toxins.[26]
- **Eggshell Contamination**: Some studies have begun exploring the presence of microplastics in reptile nesting grounds, which may reduce hatching success by altering thermal or moisture conditions.[27]

Protecting amphibians and reptiles from microplastic exposure is critical, not just for biodiversity conservation, but also for controlling pests and maintaining ecosystem balance.[28]

Microplastics and Ecosystem Services

Ecosystem services are the natural processes and outputs that support human life and economic systems—clean water, fertile soil, pollination, climate regulation, and more. Nano and microplastics have the potential to undermine these essential services by interfering with the organisms and processes that sustain them.

Pollination and Plant Reproduction

Pollinators such as bees, butterflies, and beetles are vulnerable to airborne microplastics and contaminated nectar sources.

- **Nesting Material Contamination**: Solitary bees have been found using microplastics to line their nests, potentially affecting larval development.[29]
- **Trophic Dilution**: If microplastics reduce the nutritional value of floral resources or prey, pollinator populations may decline.[30]
- **Floral Microplastic Deposition**: Airborne particles can settle on flowers, potentially interfering with pollen adhesion or stigma receptivity.[31]

Plant-pollinator relationships are finely tuned; disruptions may reduce plant reproductive success and lead to community-wide declines in vegetation diversity and abundance.[32]

Nutrient Cycling

Microplastics interfere with natural decomposition and nutrient cycling processes:

- **Litter Breakdown**: Soil fauna that decompose leaf litter, such as springtails and earthworms, suffer reduced performance when exposed to microplastics, slowing decomposition rates.[33]
- **Mycorrhizal Disruption**: These fungi form symbiotic relationships with plant roots, enhancing nutrient uptake. Microplastic-induced changes in soil pH or microbial competition can compromise these relationships.[34]
- **Marine Nutrient Cycling**: Plastics sinking through the water column can carry attached microbes and chemicals into deeper ocean layers, potentially disrupting bio-geochemical cycles and carbon sequestration.[35]

As these fundamental processes are impaired, soil and water quality suffer, leading to reduced agricultural productivity, lower biodiversity, and diminished ecosystem resilience.[36]

Food Web Stability

At the core of ecosystem functioning lies the food web—a dynamic network of producer-consumer relationships that maintain energy flow and population control. Nano and microplastics destabilize these webs in several ways:

- **Bottom-up effects**: When foundational species (like phytoplankton, zooplankton, or detritivores) decline due to plastic ingestion or toxicity, predators up the chain experience food scarcity, reduced growth, and lower reproduction.[37]
- **Top-down effects**: Apex predators affected by bioaccumulated plastic toxins can decrease predation pressure, allowing lower trophic species to proliferate unchecked. This may lead to overgrazing, algal blooms, or imbalances in community structure.[38]
- **Cross-boundary contamination**: Migratory species such as salmon, whales, and birds move plastics between ecosystems (e.g., ocean to river, forest to farmland), spreading contamination across biomes.[39]

Ultimately, food web disruptions decrease biodiversity, reduce resource availability, and increase vulnerability to invasive species and environmental stressors like drought or climate change.[40]

Case Studies: Real-World Examples of Plastic Impact on Ecosystems

To contextualize the scientific insights and ecological concepts discussed so far, it is helpful to examine real-world case studies that illustrate the tangible consequences of nano and microplastic pollution in diverse ecosystems.

Case Study 1: The Great Pacific Garbage Patch and Pelagic Food Chains

The Great Pacific Garbage Patch, a massive accumulation of marine debris located between Hawaii and California, is often cited as the most visible symbol of oceanic plastic pollution. While dominated by macroplastics like fishing nets and containers, the area also harbors high concentrations of microplastics suspended in the upper ocean.

- **Zooplankton ingestion**: Studies have documented high levels of microplastic ingestion by copepods and other zooplankton species, which are foundational to the marine food web.[41]
- **Fish contamination**: Small fish such as lanternfish, a key prey species for tuna and seabirds, are frequently found with microplastics in their digestive tracts.[42]
- **Predator exposure**: Larger predators, including tuna, sharks, and seabirds like albatrosses, accumulate plastics through the food chain. This bioaccumulation has been linked to liver damage and reproductive failure.[43]

This case study highlights how localized plastic pollution can have far-reaching effects across multiple trophic levels, impacting global fisheries and migratory bird populations.[44]

Case Study 2: River Thames and Urban Freshwater Biodiversity

One of the most studied urban rivers in the world, the River Thames in the United Kingdom serves as a prime example of how nano and microplastics infiltrate freshwater ecosystems.

- **Microplastic hotspots**: Surveys have found up to 94,000 microplastic particles per square meter of sediment in the Thames, particularly near urban outflows and combined sewer overflows.[45]
- **Benthic invertebrates**: Insect larvae, mollusks, and oligochaete worms show evidence of plastic ingestion, affecting mobility, development, and reproductive cycles.[46]
- **Birds and mammals**: Waterfowl and otters feeding in and around the riverbanks are also exposed, with reported changes in feeding behavior and gut health.[47]

The Thames illustrates the intimate connection between urban pollution, wildlife health, and ecological integrity in freshwater systems.

Case Study 3: Agricultural Plastics in China's Farmlands

Plastic mulch is widely used in agriculture to conserve water and suppress weeds. In China, where usage is especially prevalent, studies have shown that extensive use has resulted in the buildup of plastic residues in soils.

- **Soil degradation**: Residual plastics alter soil structure, reduce permeability, and interfere with microbial processes crucial for crop growth.[48]

- **Crop performance**: Reduced root development and nutrient uptake have been linked to long-term plastic accumulation.[49]
- **Food chain entry**: Crops grown in contaminated soils can absorb plastic-derived chemicals, posing dietary risks to humans and animals.[50]

This case demonstrates how agricultural practices can become a direct source of plastic contamination, threatening food security and ecosystem services.

Conservation and Remediation Efforts

Despite the severity of the crisis, global efforts are underway to mitigate the ecological damage caused by nano and microplastics. These efforts span from grassroots initiatives to international policy measures.

Habitat Cleanup and Restoration

- **Beach cleanups**: Community-led efforts in countries like Indonesia, the Philippines, and the United States help reduce macroplastics, preventing them from degrading into microplastics.[51]
- **River barriers and skimmers**: Devices like The Ocean Cleanup's Interceptor are being deployed to capture plastics in rivers before they reach the ocean.[52]
- **Sediment remediation**: Some pilot projects have experimented with filtering or extracting plastics from contaminated sediments, though scaling these efforts remains a challenge.[53]

While not a complete solution, physical removal helps reduce ongoing exposure and damage to local wildlife.

Bioremediation Research

Bioremediation uses living organisms—often microbes or fungi—to break down pollutants. Recent studies suggest certain bacterial strains can metabolize plastics like polyethylene or polystyrene under specific conditions.

- **Enzymatic breakdown**: Enzymes such as PETase and MHETase are being studied for their potential to degrade polyethylene terephthalate (PET) into harmless byproducts.[54]
- **Fungal applications**: Fungi like *Aspergillus* and *Penicillium* have shown promise in breaking down plastic polymers in lab environments.[55]
- **Synthetic biology**: Genetically engineered microbes may be developed to accelerate decomposition in landfills or composting facilities.[56]

While still experimental, these methods offer long-term hope for reducing the environmental persistence of plastics.

Policy and Protected Areas

- **Marine Protected Areas (MPAs)**: Nations have established MPAs to restrict fishing, shipping, and industrial activities, reducing the introduction of plastics into critical habitats.[57]
- **Bans on single-use plastics**: Many countries have enacted partial or full bans on plastic bags, straws, and microbeads to curb the influx of primary microplastics.[58]

- **International agreements**: Treaties like the Basel Convention and the proposed Global Plastics Treaty aim to regulate trans-boundary plastic waste and promote circular economy principles.[59]

Legislative and policy changes are critical levers for addressing the root causes of plastic pollution and protecting biodiversity.

Final Thoughts

The ecological and biodiversity impacts of nano and microplastics are profound, diverse, and global in scope. These particles infiltrate every ecosystem, from the most remote Arctic ice shelf to bustling city parks, disrupting life processes at every scale. The invisible nature of the threat makes it even more insidious—damaging organisms and habitats often before we are aware of its presence.

Ecosystems operate on delicate balances. When keystone species are disrupted, or foundational interactions like pollination, decomposition, and predation are impaired, the resilience of the entire system erodes. Nano and microplastics are not just litter—they are ecological stressors that undermine the natural infrastructure on which life depends.

This chapter has shown how urgent the problem is, but also how targeted intervention, scientific innovation, and community engagement can slow and potentially reverse the damage. The fight against plastic pollution is one of the defining environmental challenges of our time—and one that demands a coordinated, informed, and sustained global response.

Chapter 8: Hotspots and Safe Havens

Introduction

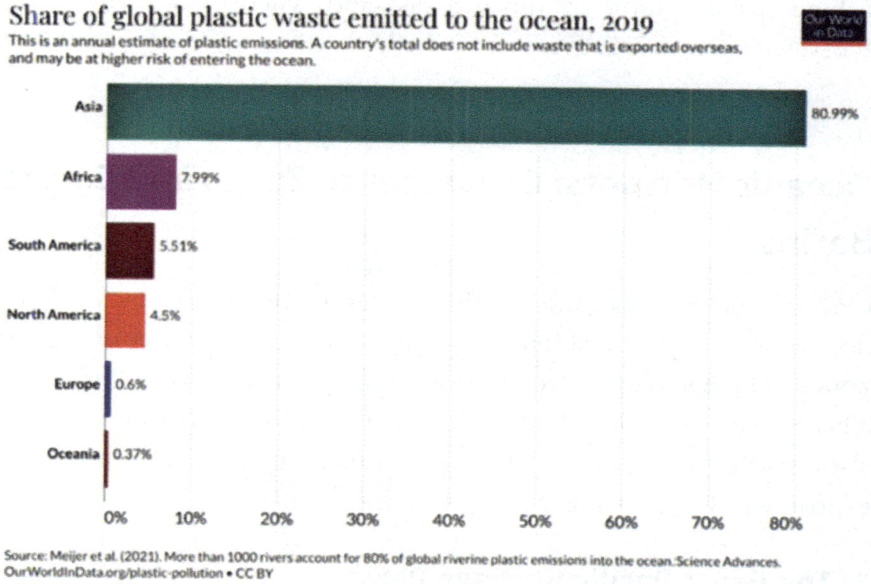

Figure 11: Global Plastic Waste Emitted to the Ocean
Source: Our World in Data, CC BY 4.0 <https://creativecommons.org/licenses/by/4.0>, via Wikimedia Commons

The global spread of nano and microplastics has reached virtually every corner of the Earth. Yet, the concentration of these pollutants is far from uniform. Certain regions act as **hotspots**, where plastic pollution accumulates to extreme levels due to geography, human activity, or environmental dynamics. Others, in contrast, still retain a degree of protection, functioning as **safe havens** where plastic intrusion is minimal or at least mitigated. Mapping these zones is essential for understanding risk, setting policy priorities, and deploying cleanup or conservation resources effectively.

This chapter explores how nano and microplastic pollution is distributed globally across water, soil, air, and biological systems. We'll identify regions suffering from high contamination, the mechanisms behind accumulation in those areas, and where—and why—some ecosystems remain relatively untouched.

Oceanic Hotspots: Convergence Zones and Coastal Basins

One of the most significant discoveries in oceanography in recent decades is the accumulation of floating plastic debris in **convergence zones**, where surface currents concentrate particles into massive gyres. These zones are not only contaminated with visible plastic waste but also serve as incubators for micro- and nanoplastics due to prolonged exposure to UV radiation and wave action.

1. The Great Pacific Garbage Patch

Located between Hawaii and California, this massive area of ocean is one of the most studied plastic hotspots. Though its name suggests a visible garbage heap, it's more accurately a **plastic soup**—containing a dense mixture of large and microscopic plastic particles suspended in the water column.

- Microplastic concentration in surface waters here is estimated at over **1.8 trillion pieces.**[1]
- High residence time in the gyre leads to extensive fragmentation and chemical leaching.[2]

- Marine life such as jellyfish, lanternfish, and sea turtles show high rates of plastic ingestion in this area.[3]

2. Other Gyres and Coastal Accumulations

The Atlantic, Indian, and Southern Oceans all have their own garbage patches or **subtropical gyres**, albeit less studied than their Pacific counterpart. Coastal areas near river mouths also accumulate plastics due to discharge from urban and agricultural runoff.

- The Bay of Bengal and the North Sea have been recorded as **high-density plastic zones.**[4]
- Large rivers like the Yangtze, Ganges, and Amazon serve as major transporters of microplastics from land to sea.[5]
- Estuaries and deltas become sedimentation zones for both floating and sinking plastic particles.[6]

Freshwater Hotspots: Rivers, Lakes, and Urban Runoff

While oceans attract much attention, **freshwater systems** are critical conduits for plastic pollution. These ecosystems receive plastic waste directly from cities, agriculture, and industry before it reaches marine environments.

1. Highly Contaminated Rivers

A 2017 study by the Helmholtz Centre for Environmental Research found that **ten rivers** are responsible for over **90% of plastic waste** entering the oceans.

- These include the Yangtze, Indus, Yellow, Hai He, Nile, Ganges, Pearl, Amur, Niger, and Mekong Rivers.[7]
- Many of these flow through densely populated or industrialized regions with inadequate waste management.[8]
- High plastic loads have been found in sediments and fish tissues along these rivers, including nano-sized particles in drinking water samples.[9]

2. Polluted Lakes and Reservoirs

Lakes can act as both **sinks and sources** of plastic pollution depending on their hydrology and human influence.

- The **Great Lakes** in North America, especially Lake Erie and Lake Ontario, contain high concentrations of microplastics from urban discharge and industrial waste.[10]
- In Europe, **Lake Geneva** and **Lake Constance** have reported measurable nanoplastic particles, likely introduced through wastewater treatment plant outflows.[11]
- Floating plastics can interfere with light penetration and phytoplankton growth, affecting entire aquatic food webs.[12]

These regions are especially vulnerable because their closed or semi-closed hydrological systems prevent rapid flushing of contaminants, resulting in accumulation over time.

Soil-Based Hotspots: Agricultural Fields, Landfills, and Biosolid Sites

While attention often focuses on aquatic pollution, **soils are increasingly recognized as reservoirs for microplastics**. Due to the weight and physical properties of many plastic particles, they tend to settle rather than remain airborne or aquatic, accumulating in terrestrial environments over time.

1. Agricultural Soils and Plastic Mulching

Plastic mulch is widely used in agriculture to suppress weeds and retain moisture. Over time, however, it breaks down into smaller fragments and contributes to a growing microplastic load in soil.

- In regions like **Northwest China**, where plasti-culture is heavily adopted, soil samples have revealed microplastic concentrations exceeding **4,000 particles per kilogram**.[13]
- These plastics change the soil's porosity, affect microbial diversity, and reduce crop productivity over multiple planting cycles.[14]
- Chemical additives such as UV stabilizers and flame retardants leach into the soil and may enter crops through root systems.[15]

Plastic mulching is also common in parts of India, Israel, and the western United States, making these regions potential **hotspots for terrestrial microplastic accumulation**.

2. Wastewater Sludge and Biosolids

Treated sewage sludge, often referred to as **biosolids**, is commonly used as a fertilizer. Unfortunately, wastewater treatment plants do not fully remove microplastics from domestic and industrial effluent.

- An estimated **90% of microplastics** entering wastewater plants are retained in sludge.[16]
- When this sludge is applied to farmland, it becomes a direct pipeline for plastic contamination.[17]
- The UK alone applies over **3.5 million tons of biosolids annually**, with significant microplastic loads embedded in the waste.[18]

This mechanism creates long-term reservoirs of plastic pollution in agricultural soils and forested lands, with unknown consequences for soil health and food safety.

3. Landfills and Dump Sites

Landfills are perhaps the most intuitive source of microplastic contamination, but their contribution extends beyond just buried waste.

- Plastics degrade slowly under environmental exposure, especially when landfills are unlined or poorly managed.[19]
- Wind-blown dust and leachate from these sites introduce plastic particles into neighboring fields, groundwater, and even air.[20]
- In many low-income regions, open dumps are common, where plastic is burned or scavenged—releasing both particulate pollution and toxic residues.[21]

Landfill-adjacent areas in countries like Indonesia, Nigeria, and parts of Latin America show high levels of microplastics in both **surface soils and groundwater**, representing long-term contamination hotspots.

Atmospheric Hotspots: Airborne Plastics in Cities and Mountains

Less visible but equally concerning is the presence of **microplastics and nanoplastics in the atmosphere**, transported by wind and eventually deposited in remote or urban environments.

1. Urban Air Pollution and Indoor Environments

Cities are significant sources of airborne plastic particles. Tire wear, synthetic textile fibers, construction dust, and plastic waste all contribute to the atmospheric load.

- A 2020 study found that urban air in Paris, Hamburg, and London contains **between 50 and 400 microplastic particles per cubic meter.**[22]
- Indoor air often contains even higher concentrations due to carpet abrasion, clothing fibers, and dust from plastic-containing furniture.[23]
- People may inhale tens of thousands of microplastic particles annually, depending on lifestyle and exposure.[24]

These cities serve as "airborne microplastic hotspots," particularly around transportation corridors and high-density housing.

2. Remote Deposition in Mountains and Polar Regions

What surprised researchers most in recent years is the discovery that **airborne microplastics are traveling thousands of kilometers**, settling even in remote locations.

- The Pyrenees Mountains, Arctic snowfields, and Swiss Alps have all been found to contain microplastics in snow and surface air.[25]
- Researchers estimate that particles can be lofted into the upper atmosphere, traveling on wind currents before settling.[26]
- These regions were once believed to be pristine, yet their contamination suggests global reach of airborne plastics.[27]

The presence of plastic particles in areas without direct human activity underscores how far-reaching this pollution has become, defying assumptions of geographic safety.

Summary So Far

From **agricultural lands** soaked in biosolid fertilizers to **urban skies** choked with synthetic dust, the story of nano and microplastics is one of astonishing infiltration. Even our most remote mountaintops and farmlands—often imagined as untouched sanctuaries—have become part of the plastic pollution narrative.

Food Chain Contamination Hotspots: Where We Eat What We Pollute

One of the most alarming implications of nano and microplastic pollution is its entry into the **global food web**. Contamination is no longer limited to remote rivers or distant gyres—these plastics are now being detected in food items routinely consumed across the world. Certain regions, due to dietary habits, environmental exposure, or food processing techniques, emerge as **hotspots for food chain contamination**.

1. Seafood-Rich Regions

Populations that rely heavily on seafood consumption, particularly **filter-feeding organisms**, are at greater risk of dietary plastic exposure.

- Studies in Japan, South Korea, and coastal China found microplastics in **over 70% of mussels, clams, and oysters** sampled from local markets.[28]
- Filter feeders ingest suspended particles indiscriminately, making them high-risk vectors for human exposure.[29]
- Ingested plastics in shellfish are not easily removed by rinsing or cooking, as the entire organism is often consumed.[30]

Coastal fishing communities in Southeast Asia, the Mediterranean, and the Pacific Islands frequently consume **whole small fish and mollusks**, increasing exposure levels compared to Western consumers who mostly eat fillets.

2. Agricultural Produce and Processed Foods

Plastics can infiltrate **plant-based food** systems through irrigation, atmospheric fallout, and contaminated fertilizers.

- Vegetables grown in soil treated with biosolids show trace plastic particles in root and leafy tissues.[31]
- Processed food items such as salt, sugar, rice, and tea bags have tested positive for microplastic contamination.[32]
- Sea salt, especially from regions near plastic accumulation zones like the **South China Sea**, can contain **up to 600 particles per kilogram.**[33]

Even packaging contributes to contamination: plastic wrappings, containers, and processing tools can shed particles that mix with the food, especially during cutting, sealing, or heating.

3. Bottled Water and Beverages

Among the most direct forms of ingestion is via **drinking water,** especially bottled products.

- A global study by Orb Media in 2018 found that **93% of bottled water** samples contained microplastic particles.[34]
- Brands from the U.S., India, Indonesia, Mexico, and Brazil were tested, with some samples showing **over 10,000 particles per liter.**[35]
- The contamination originates from bottling processes, bottle caps, and possibly the water source itself.[36]

This turns everyday hydration into a source of chronic plastic exposure, especially in areas where bottled water is consumed for lack of clean tap alternatives.

4. Dairy, Meat, and Eggs

Animal products are also affected, especially when livestock are raised in environments where plastic particles are present in feed, water, or air.

- Chickens raised near urban dump sites have been found to contain microplastics in their **gizzards and muscle tissues.**[37]
- Cow's milk from certain dairy farms in Germany and India has shown trace levels of synthetic particles.[38]
- The contamination is thought to originate from feed additives, airborne dust, and mechanical equipment made of plastic.[39]

While more research is needed, these findings point to a **systemic integration of plastics into terrestrial food chains**, not just marine ones.

Medical and Industrial Hotspots

Beyond environmental and dietary exposure, certain occupational and medical environments also represent **microplastic exposure hotspots** due to the materials and processes used.

1. Hospitals and Medical Facilities

Healthcare settings make heavy use of single-use plastics—tubing, gloves, masks, syringes, and packaging—all of which contribute to particulate pollution in indoor air and surfaces.

- Intubation, nebulization, and high-speed drilling procedures may introduce plastic particles into airways or wounds.[40]

- Neonatal incubators, dialysis equipment, and IV systems use plastic components that can shed nano- and microplastics.[41]
- Hospital waste often escapes rigorous filtration during incineration or disposal, especially in low-resource settings.[42]

This makes certain healthcare contexts inadvertent sources of **microplastic exposure to patients and staff alike**.

2. Industrial Facilities

Textile factories, plastic molding units, and tire manufacturing plants release a wide range of synthetic particles into surrounding air, water, and soil.

- Workers in these environments often report respiratory issues, and studies have detected microplastic particles in nasal mucus and lung tissue.[43]
- Surrounding communities may suffer long-term exposure, as airborne plastics settle on homes, food crops, and water sources.[44]
- Environmental monitoring near industrial corridors in **Bangladesh, Turkey, and parts of the U.S. Midwest** confirms elevated microplastic levels in nearby ecosystems.[45]

The convergence of occupational hazards and environmental contamination highlights how plastic pollution operates not just ecologically, but also socioeconomically.

Safe Havens: Where Are Microplastic Levels Low?

Although plastic contamination has reached virtually every corner of the planet, there are still **relative safe havens**—regions or micro-environments where nano and microplastic concentrations are significantly lower. These areas are valuable for research baselines, ecological conservation, and public health guidance.

1. Remote Forest Ecosystems

Dense, undisturbed forests—such as the Amazon Rainforest, Congo Basin, and boreal forests of Canada and Russia—tend to exhibit **lower microplastic loads** than agricultural or urban-adjacent areas.

- The canopy provides a natural buffer against atmospheric deposition of airborne particles.[46]
- Lack of intensive human presence minimizes direct plastic input from waste, construction, or vehicular traffic.[47]
- Soil samples from protected zones in the Amazon reveal microplastic concentrations under **50 particles per kilogram**, a fraction of what's found in farmed or industrial soil.[48]

These ecosystems are critical for biodiversity and climate regulation and offer a window into what "pre-plastic" environmental baselines might have looked like.

2. Deep Ocean Sediments

Ironically, while surface waters are teeming with plastic particles, **deep-sea sediments** in certain regions remain less affected.

- At depths greater than 6,000 meters, such as the Mariana Trench or certain abyssal plains, plastic deposition is slowed by currents and limited vertical mixing.[49]
- Sampling of these zones has shown particle concentrations to be **orders of magnitude lower** than continental shelf or coastal zones.[50]
- The remoteness and high pressure may delay degradation, although long-term accumulation remains a risk.[51]

Researchers propose that some deep-sea regions be treated as **sentinel sites**—global benchmarks to track how fast and far plastic contamination spreads.

3. Indigenous Lands and Protected Reserves

Lands stewarded by Indigenous communities, or protected under government or international agreements, often show **lower plastic contamination levels** due to lower development intensity.

- Areas managed by Indigenous groups in Canada, Australia, and the Amazon typically have reduced industrial activity, minimized chemical inputs, and less reliance on plastic-packaged goods.[52]
- Marine protected areas (MPAs) with restrictions on fishing, tourism, and waste discharge often exhibit **lower concentrations of plastics in water, sand, and marine life**.[53]
- Examples include the Papahānaumokuākea Marine National Monument in Hawaii and the Chagos Archipelago in the Indian Ocean.[54]

Empowering local custodianship of land and water remains a key strategy for maintaining and replicating these safer zones globally.

4. Certain Freshwater Springs and Glacial Sources

Because groundwater and glacial meltwater are typically **filtered naturally**, they often emerge as low-contaminant sources—at least initially.

- Alpine glacial springs in Switzerland, high-altitude Andean lakes, and Icelandic lava-filtered springs show **minimal microplastic content** upon exit from natural reservoirs.[55]
- However, contamination can begin immediately downstream via airborne deposition or contact with plastic pipes and containers.[56]
- Preservation of these "cold start" points is essential for long-term water quality monitoring.[57]

These rare clean sources serve as **control groups** in environmental monitoring and can inform purification benchmarks for drinking water systems.

The Importance of Hotspot Mapping and Monitoring

Identifying both **hotspots and safe havens** is critical to crafting effective responses to nano and microplastic pollution. This requires **standardized monitoring** and **spatial mapping tools**.

- Global efforts like the **UNEP's World Environment Situation Room** and **European Space Agency's plastic tracking satellites** are improving geographic data resolution.[58]

- Citizen science programs, such as the **Adrift project** and **Plastic Pollution Tracker**, enable crowd-sourced data from kayakers, hikers, and divers.[59]
- Scientists use sediment cores, air samplers, and biological tissue analysis to create time-stamped contamination records.[60]

By understanding where the problem is most acute—and where it is least—policymakers can prioritize interventions, fund clean-up efforts, and preserve refuges.

Final Thoughts on Geographic Disparities

The uneven spread of nano and microplastic pollution illustrates the **global imbalance** of environmental burdens. Developing countries often bear the brunt of landfill overflow and illegal dumping, while industrialized nations export waste or consume at disproportionate rates.

Even so, **no region is truly safe**. The illusion of pristine nature has been pierced by data showing microplastics in rainfall, Antarctic snow, and placental tissues. Hotspots demand urgency; safe havens demand protection.

Solutions must be targeted, scalable, and informed by rigorous monitoring. Whether we live on a tropical island, in a sprawling metropolis, or high in the mountains, the challenge of microplastic pollution is universal—but not uniform.

The next chapter will shift focus from where the plastics are to what individuals can do about them. Detox starts at home, but echoes outward.

Chapter 9: Strategies for Personal Detox

Introduction

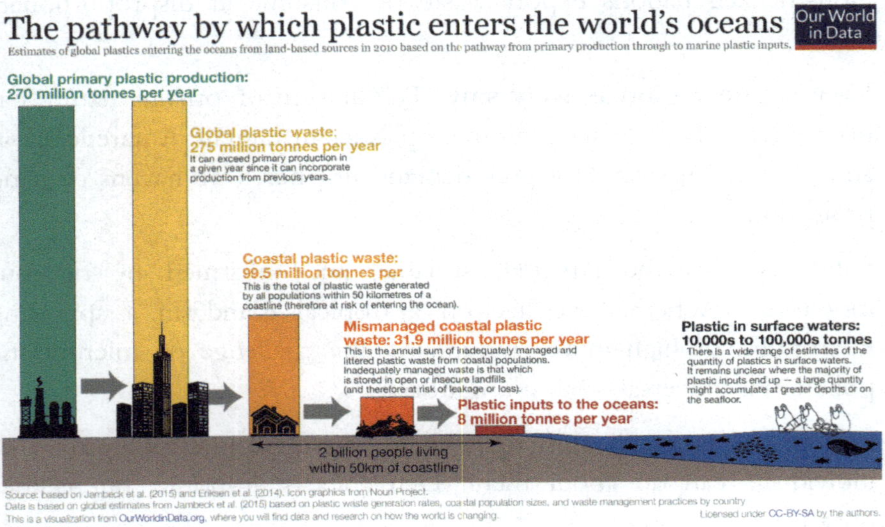

Figure 12: Pathway by Which Plastic Enters Oceans
Source: Our World in Data, CC BY 4.0 <https://creativecommons.org/licenses/by/4.0>, via Wikimedia Commons

Plastic is no longer just a visible pollutant cluttering landscapes or swirling in ocean gyres—it now exists invisibly, pervasively, and disturbingly within the human body. Nano and microplastics, the microscopic fragments that result from the breakdown of larger plastic waste or are manufactured at small sizes, have been detected in human blood, lungs, placenta, and fecal matter.[1] They enter the body silently, often unnoticed and undetected, but the long-term health effects are only beginning to emerge. While societal-level solutions—such as international regulations and corporate reform—are critical to solving the global plastic problem, individuals are not helpless. You have the ability to take action, starting within your own home, habits, and body.

This chapter provides a comprehensive guide to detoxifying your life and body from nano and microplastics. It focuses on practical, evidence-based strategies to minimize intake, enhance your natural detox systems, and reduce exposure in everyday life. These strategies don't promise to eliminate all plastics from your environment—that would be nearly impossible in the modern world—but they do equip you to reduce your body burden and promote resilience in a world saturated with synthetic particles.

Understanding How Plastics Enter the Body

The first step in any detox protocol is understanding the pathways through which the toxin—here, nano and microplastics—enters the body. These particles can infiltrate through several major routes:

- **Ingestion**: Plastics are now common in our food and water supply. Studies show they are present in seafood, table salt, honey, fruits, vegetables, tea bags, and even beer.[2] Bottled water is a major contributor.[3] Packaging materials, plastic kitchenware, and food processing equipment can all leach plastic residues into food.
- **Inhalation**: Indoor air, particularly in homes with synthetic textiles and poor ventilation, can be rich in airborne microplastic fibers.[4] These are released from carpets, furniture, clothing, and dust. Vacuuming and air movement stir them into the air, where they can be inhaled deep into the lungs.
- **Dermal absorption**: Though the skin is a relatively strong barrier, nano-scale plastics can potentially pass through the skin,

particularly if used in personal care products such as exfoliants or lotions.[5] Plasticizers and solvents that accompany microplastics may be absorbed even more easily.

- **Medical exposure**: Intravenous fluids and pharmaceuticals packaged in plastics can introduce trace amounts directly into the bloodstream, though this is still under study.[6]

Step 1: Source Reduction – Stop the Flow

Before detoxifying what's already in the body, it's essential to stop ongoing exposure. Reducing plastic exposure at its source is the most powerful and effective detox strategy.

Food and Beverage Safety

- **Filter your water**: Tap water may contain plastic fibers, especially in urban areas. Use a certified reverse osmosis or activated carbon filtration system to reduce plastic contamination.[7]
- **Avoid bottled water**: Bottled water has been shown to contain twice as many microplastic particles as tap water.[8] Stick to filtered water in stainless steel or glass containers.
- **Minimize plastic food packaging**: Buy food in bulk using cloth or paper bags. Choose products stored in glass jars or cardboard containers when possible.[9]
- **Never microwave food in plastic**: Heating plastic containers can accelerate the breakdown and release of microplastics and endocrine-disrupting chemicals like BPA and phthalates into food.[10]

- **Avoid non-stick cookware with plastic coatings**: Teflon and other non-stick surfaces can flake off during cooking. Choose cast iron, stainless steel, or ceramic cookware.[11]

Clothing and Household Textiles

- **Choose natural fibers**: Clothing made from cotton, wool, hemp, or linen sheds fewer synthetic fibers. Polyester, nylon, acrylic, and elastane contribute heavily to indoor microfiber pollution.[12]
- **Use a microfiber-catching laundry bag or filter**: Products like the Guppyfriend Washing Bag or external filters can capture microfibers released during washing.[13]
- **Dry clothes outside**: Tumble dryers can break down fibers further. Sun drying reduces microfiber shedding.[14]
- **Vacuum with HEPA filters**: Regular vacuuming with a HEPA filter can reduce the number of airborne particles in your home.[15]

Kitchen and Household Products

- **Replace plastic utensils and containers**: Wooden, bamboo, and metal kitchen tools are safer options. Switch plastic cutting boards for wooden ones.[16]
- **Avoid plastic wrap and cling film**: Beeswax wraps, silicone lids, and glass containers make excellent reusable alternatives.[17]
- **Use non-plastic dish-washing tools**: Loofahs, metal scrubbers, or biodegradable sponges can replace synthetic dish pads.[18]

- **Clean smart**: Use vinegar, baking soda, and essential oils for cleaning instead of chemical-laden products in plastic spray bottles.[19]

Step 2: Supporting the Body's Natural Detox Systems

The body is equipped with remarkable detoxification systems: the liver, kidneys, gastrointestinal tract, lungs, skin, and lymphatic system. Optimizing these systems helps the body clear out plastic-related compounds and maintain resilience.

Liver Support

The liver metabolizes toxins and prepares them for excretion. To support its function:

- **Eat cruciferous vegetables**: Broccoli, cauliflower, kale, and Brussels sprouts contain glucosinolates that help activate detox enzymes.[20]
- **Supplement with milk thistle or dandelion root**: Both support liver regeneration and bile production.[21]
- **Avoid alcohol and ultra-processed foods**: These burden the liver and reduce its detox efficiency.[22]

Kidney Function

The kidneys excrete water-soluble toxins via urine. To assist them:

- **Stay hydrated**: Drink clean, filtered water—at least 2 to 3 liters per day.[23]

- **Eat potassium-rich foods**: Bananas, sweet potatoes, and avocados support electrolyte balance.[24]
- **Avoid excessive NSAID use**: Overuse of ibuprofen or naproxen can harm the kidneys over time.[25]

Gut Health

The gastrointestinal tract not only digests food but also traps and excretes harmful substances.

- **Eat prebiotics and probiotics**: Fiber from oats, garlic, and legumes feeds healthy gut flora. Yogurt, kefir, and fermented vegetables replenish good bacteria.[26]
- **Add soluble fiber**: Flaxseed, psyllium husk, and chia seeds bind toxins and promote regularity.[27]
- **Avoid constipation**: Regular elimination is crucial for toxin clearance.[28]

Skin and Lymphatic Detox

The skin is a major organ of detoxification, especially through sweat, and the lymphatic system helps clear waste and toxins from body tissues. Enhancing these systems can promote the removal of plastic-related residues and synthetic contaminants.

- **Sweating and sauna use**: Infrared saunas are increasingly studied for their role in detoxification. Regular sessions (2–3 times per week) may support the excretion of chemical toxins through sweat.[29] Be sure to hydrate properly and consult a physician, especially if you have cardiovascular issues.

- **Dry brushing**: This ancient technique stimulates lymph flow and exfoliates dead skin. Use a firm-bristled natural brush and brush toward the heart in circular motions before showering.[30]
- **Epsom salt baths:** These supply magnesium sulfate, which can assist muscle relaxation and possibly aid lymphatic drainage. Add 1–2 cups to warm water and soak for 20 minutes.[31]

Respiratory Detox

Because microplastic fibers are inhaled through the air, particularly indoors, it's important to protect and cleanse the lungs:

- **Breathe clean air**: Use HEPA air purifiers, avoid smoking, and reduce exposure to incense and candles that produce soot or fine particles.[32]
- **Practice deep breathing exercises**: Diaphragmatic breathing and yogic pranayama exercises improve lung capacity and may help expel pollutants more effectively.[33]
- **Incorporate lung-supporting herbs**: Mullein, elecampane, and licorice root have traditional use in respiratory health. Use teas or supplements under professional guidance.[34]

Step 3: Nutritional Approaches to Microplastic Detox

Diet is not just about avoidance—it's also about empowerment. Certain nutrients and compounds can support your body's ability to deal with plastic-related chemicals, oxidative stress, and inflammation.

Antioxidants

Antioxidants help neutralize free radicals, which are often generated in response to foreign particles like plastic residues.

- **Vitamin C**: Found in citrus fruits, bell peppers, strawberries, and broccoli. Enhances glutathione production, a key antioxidant.[35]
- **Vitamin E**: Present in almonds, sunflower seeds, and spinach. Protects cell membranes from lipid peroxidation.[36]
- **Polyphenols**: Found in green tea, turmeric, dark chocolate, and berries. Quercetin and curcumin in particular have potent anti-inflammatory effects.[37]

Fiber and Binding Agents

Soluble fibers and certain natural binders can help capture toxins in the digestive tract and aid in their elimination.

- **Flaxseed and chia seeds**: Provide mucilage and bind toxins while supplying omega-3 fatty acids.[38]
- **Activated charcoal**: Occasionally used in detox protocols to trap chemicals, but should be used short-term and away from medication or nutrient intake.[39]
- **Chlorella and spirulina**: Algae rich in chlorophyll, which may help bind heavy metals and synthetic compounds, though data on plastics is limited.[40]

Hormone Support

Since many plasticizers (like phthalates and BPA) act as endocrine disruptors, supporting hormonal balance is a key part of detox.

- **Cruciferous vegetables**: Aid in estrogen metabolism through the compound DIM (diindolylmethane).[41]
- **Healthy fats:** Eat avocado, olive oil, and wild-caught salmon to support hormone production and cell membrane health.[42]
- **Liver support**: As discussed earlier, estrogen and synthetic xenoestrogens are processed by the liver.[43]

Step 4: Behavioral and Lifestyle Modifications

Small, consistent behavioral changes often add up to the most powerful results over time.

Conscious Consumption

- **Perform a plastic audit**: Track how much plastic you use in a day or week. This increases awareness and highlights areas to improve.[44]
- **Choose sustainable alternatives**: Opt for glass straws, bamboo toothbrushes, refillable toiletry bottles, and bar soaps.[45]
- **Avoid synthetic fragrances**: Found in candles, sprays, and lotions, these often contain phthalates and other plastic-based carriers.[46]

Food Preparation and Storage

- **Use non-toxic cookware**: Ceramic, stainless steel, and cast iron pans are ideal.[47]
- **Store food in glass containers**: Especially for oily or acidic foods, which leach plastics more easily.[48]

- **Avoid canned food**: Many cans are lined with plastic resins that can leach into food.[49]

Sleep and Stress Management

Detox processes ramp up during rest. Chronic stress, meanwhile, elevates cortisol, which can impair detox pathways.

- **Prioritize 7–9 hours of quality sleep**: Darkness, a cool room, and consistent schedules help.[50]
- **Incorporate relaxation techniques**: Meditation, journaling, and nature walks reduce stress hormones.[51]
- **Limit blue light exposure**: Avoid screens 1 hour before bed or use blue light filters.[52]

Psychological and Emotional Resilience

Facing the enormity of plastic pollution can be overwhelming. However, mindset plays a key role in maintaining motivation and avoiding burnout or eco-anxiety.

Set Realistic Goals

- **Start with small wins**: Replace one high-exposure product per week—like your shampoo or water bottle.[53]
- **Track progress visually**: Use habit trackers or apps to monitor your detox journey.[54]

Engage with Community

- **Join or form support groups**: Online forums, neighborhood workshops, or eco-activist circles can offer camaraderie.[55]

- **Celebrate milestones**: Every sustainable switch is a victory, whether it's ditching plastic wrap or filtering your tap water.[56]

Educate Others

- **Lead by example**: Your habits may inspire others to take similar steps.[57]
- **Share resources**: Recommend documentaries, articles, or books that explain the plastic problem and solutions.[58]

Step 5: Technological and Innovative Aids

Modern innovation is also helping individuals reduce plastic exposure in novel ways. These tools, while sometimes expensive, represent the cutting edge of personal environmental health.

Air and Water Purifiers

- **HEPA Filters**: High-efficiency particulate air (HEPA) filters are effective at capturing airborne microfibers in indoor environments. These can be installed in vacuum cleaners, standalone purifiers, or HVAC systems.[59]
- **Activated Carbon Filters**: Often paired with HEPA systems, they remove volatile organic compounds (VOCs) and odors, enhancing indoor air quality.[60]
- **Reverse Osmosis Water Systems**: These are among the most reliable at filtering microplastics from tap water. Though more costly than carbon filters, they also remove pharmaceutical residues and heavy metals.[61]

Smart Wearables and Test Kits

- **Wearable air monitors**: Some advanced models can detect particles as small as PM2.5, giving you real-time data on air quality.[62]
- **Home testing kits for water**: While still emerging, some labs allow you to mail in water samples for plastic analysis, helping identify contamination points in your home supply chain.[63]

Washing Machine Filters

- **External microfiber filters**: Devices like the Lint LUV-R or PlanetCare connect directly to your washing machine drain to catch synthetic fibers before they enter wastewater.[64]
- **Integrated drum filters**: High-end appliances now feature built-in microfiber catchers, though their availability is limited to certain markets.[65]

Clothing Innovations

- **Low-shed synthetics**: Some new textiles are engineered to shed fewer fibers. These include polyester blends that are tightly woven or chemically treated.[66]
- **Biodegradable fabrics**: Startups are developing plastic-free technical fabrics made from algae, cellulose, or mycelium.[67]

Case Study: A Family Detox Journey

The Nguyen family, a household of five living in Southern California, decided to dramatically reduce their plastic exposure over six months.

Motivated by their daughter's asthma and a local environmental film screening, they began making changes in phases.

- **Phase 1: Food and Water** – They purchased a reverse osmosis water system, stopped buying bottled drinks, and switched to stainless steel lunch containers.[68]
- **Phase 2: Household** – They replaced plastic cooking utensils, installed HEPA filters in bedrooms, and swapped out Teflon pans for ceramic.[69]
- **Phase 3: Personal Care** – The family gradually shifted to bar soaps, shampoo bars, bamboo toothbrushes, and homemade body scrubs.[70]
- **Phase 4: Clothing and Laundry** – They invested in a microfiber filter for their washer and committed to buying only organic cotton or secondhand clothes.[71]

Results after six months included reduced household waste (two garbage bags per month instead of five), improved air quality (fewer allergy symptoms), and a sense of shared purpose among family members.[72]

Research Insights: Measuring the Unseen

Recent studies have revealed the hidden extent of plastic exposure:

- **Placental plastics**: A 2020 Italian study identified microplastic fragments in placental tissue, raising concerns about fetal development.[73]
- **Infant exposure**: Research shows that infants fed with polypropylene bottles may ingest over 1 million microplastic particles per day.[74]

- **Fecal samples**: Global sampling of human feces has shown widespread microplastic presence, with average counts of 20 particles per 10 grams.[75]

While causality between microplastics and disease is still under investigation, the correlation between exposure and oxidative stress, immune response, and endocrine disruption is growing stronger in scientific literature.[76]

Taking Action: A Detox Checklist

To translate this chapter's insights into action, consider the following checklist:

- ✅ Use glass or stainless-steel containers for all food and drink
- ✅ Avoid bottled water; filter tap water instead
- ✅ Replace synthetic clothing with natural fiber alternatives
- ✅ Install a microfiber filter on your washing machine
- ✅ Choose bar soaps and avoid products with "poly" ingredients
- ✅ Use HEPA-filter air purifiers in key rooms
- ✅ Cook with ceramic or stainless-steel cookware
- ✅ Eat high-fiber and antioxidant-rich foods
- ✅ Practice dry brushing and occasional sauna use
- ✅ Engage in regular deep breathing or exercise
- ✅ Track your progress and share your journey with others [77]

Final Reflections

Detoxing from nano and microplastics is no longer a fringe concern—it is a practical necessity in a world overrun by synthetic pollution. While you cannot remove every plastic molecule from your environment, you can take meaningful steps to lower your personal burden. Each change—no matter how small—moves the needle toward better health and environmental stewardship.

Detox is not just about physical elimination. It is a form of resistance against the unchecked proliferation of disposable culture. It is a declaration that your health, and the health of your family and planet, matters more than convenience or cost-cutting corporate policies.

No government or regulation can match the immediacy of personal action. And when millions take those actions together, systemic change becomes not only possible—but inevitable.[78]

Chapter 10: Strategies for Environmental Detox

Introduction

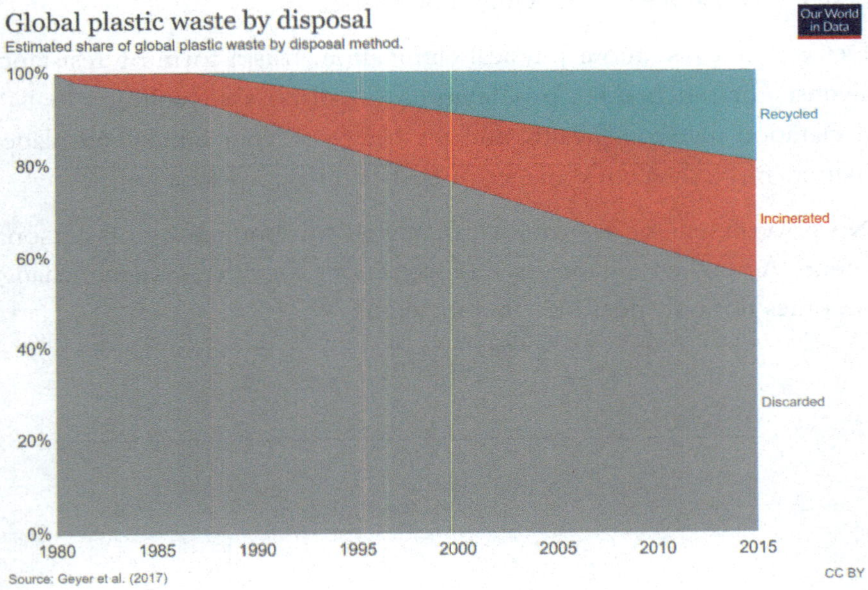

Figure 13: Global Plastic Waste by Disposal
Source: Jenna R. Jambeck, Roland Geyer, Chris Wilcox, Theodore R. Siegler, Miriam Perryman, Anthony Andrady, Ramani Narayan, Kara Lavender Law, CC BY 4.0 <https://creativecommons.org/licenses/by/4.0>, via Wikimedia Commons

As the world confronts the widespread infiltration of nano and microplastics into ecosystems, food chains, and human bodies, the need for environmental detoxification has become more urgent than ever. While cleanup alone cannot solve the plastic crisis, it plays a critical role alongside reduction and prevention. Environmental detox refers to the deliberate, collective efforts aimed at identifying, capturing, and eliminating existing plastic particles from land, water, air, and biotic systems. This chapter focuses on both grassroots and large-scale strategies for detoxifying the environment from these persistent pollutants, integrating science, technology, community action, and policy mechanisms.[1]

Understanding the Scope of Environmental Contamination

Plastic particles are not evenly distributed across the globe. Instead, they tend to accumulate in hotspots—urban centers, coastlines, agricultural soils, wastewater systems, and air ducts—yet even remote regions like the Arctic and mountain glaciers are not immune.[2] Detox strategies must therefore adapt to diverse environments, including:

- **Aquatic systems**: Rivers, lakes, oceans, wetlands.
- **Terrestrial zones**: Farmlands, forests, cities, dumpsites.
- **Atmosphere**: Indoor air, industrial emissions, particulate matter in urban settings.
- **Built environment**: Infrastructure like buildings, highways, and storm drains.

The first step to environmental detox is acknowledging that the problem is layered, and each layer requires a different intervention.[3]

Strategy 1: Cleanup Operations

A. Coastal and River Cleanups

Grassroots cleanup efforts have become widespread and are often the entry point for public engagement. These initiatives focus on visible plastics but also serve as platforms for educating communities about microplastics.

- **Voluntary beach cleanups**: Organized by NGOs, schools, or local governments. While they often collect large debris, they can also include sieving sand for microplastics.[5]
- **River interceptors**: Devices like the "Interceptor" by The Ocean Cleanup project use solar-powered barriers to trap floating plastic before it reaches the ocean.[6]
- **Floating booms and skimmers**: Installed across river mouths or industrial discharge points to capture plastics and synthetic waste.[7]

B. Ocean-Based Recovery

The ocean is the final sink for much of the world's plastic waste. Detox strategies here are more challenging due to the vastness, depth, and complexity of marine systems.[8]

- **Macroplastic removal**: Organizations like 4Ocean and Ocean Voyages Institute deploy ships and drones to collect ghost nets and large plastic debris.[9]
- **Gyre-focused strategies**: Projects targeting the Great Pacific Garbage Patch attempt to create large barriers that slowly corral plastic toward a centralized collection point.[10]

- **Microplastic filtration research**: Emerging technologies such as plankton-safe sieves and microbubble currents are being developed to separate small particles from seawater.[11]

Strategy 2: Soil and Agricultural Detox

A. Regenerative Agriculture

Agricultural land is increasingly contaminated by plastic mulch films, irrigation systems, sewage sludge, and atmospheric deposition.[12]

- **Mulch film alternatives**: Replace polyethylene sheets with biodegradable films made from starch, cellulose, or polylactic acid (PLA).[13]
- **Biochar application**: This carbon-rich product, when added to soil, may bind certain contaminants and help immobilize microplastic fragments, though research is ongoing.[14]
- **Compost scrutiny**: Encourage municipal composting systems to screen for and remove plastics before application to fields.[15]

B. Land Restoration Projects

Landfills, illegal dumps, and industrial zones serve as major microplastic contributors via leachate and runoff.[16]

- **Phyto-remediation**: Use of plants such as willows and poplars to stabilize soil and potentially uptake certain pollutants—plastic inclusion in this process is being studied.[17]
- **Topsoil replacement**: In severely polluted sites, contaminated layers are excavated and replaced, though this is labor- and cost-intensive.[18]

- **Plastic audit and zoning laws**: Municipalities can identify plastic-heavy zones and rezone them for limited access or managed restoration.[19]

Strategy 3: Airborne Plastic Capture

Airborne microplastics, released from tires, textiles, construction sites, and industrial processes, contribute to atmospheric contamination, particularly in urban and indoor environments.[20]

A. Urban Planning Innovations

Cities can mitigate microplastic circulation through smarter design:

- **Green barriers**: Planting dense vegetation along roads and industrial zones helps trap airborne particulates.[21]
- **Permeable pavements**: Reduce plastic-rich runoff and trap microfibers at the surface.[22]
- **Vertical gardens and green roofs**: Function as both air purifiers and microplastic absorbers.[23]

B. Industrial Emission Controls

Plastic production and textile manufacturing are major sources of airborne plastic particles.

- **Retrofitting factories**: Require HEPA filters and scrubbers in ventilation systems to capture plastic dust.[24]
- **Regulations on textile finishing**: Implement emission caps on fiber treatment plants and enforce closed-loop systems.[25]

- **Worker protection**: Ensure proper protective gear in plastic-heavy workplaces, indirectly reducing the environmental release of contaminated waste.[26]

Strategy 4: Wastewater Treatment Upgrades

One of the primary pathways through which nano and microplastics enter the environment is via domestic and industrial wastewater. Most modern treatment plants are not designed to fully capture particles smaller than 5 millimeters—especially nanoplastics, which can pass through conventional filtration systems.[27]

A. Advanced Filtration Technologies

Several treatment innovations are now being explored and adopted to help remove plastic particulates at the source:

- **Membrane bioreactors (MBRs)**: Combine biological treatment with microfiltration or ultrafiltration membranes, significantly improving the removal of microplastics down to 0.1 μm.[28]
- **Nanofiltration systems**: Even more precise than MBRs, these filters can catch particles at the nanoscale but come with higher maintenance and energy demands.[29]
- **Electrocoagulation**: This method uses electric currents to aggregate particles, making them easier to filter. Some pilot studies show promise in binding microplastics with metal ions for removal.[30]
- **Biofilters**: Using natural or engineered microbial communities to break down organic material and trap small particles within matrix structures.[31]

B. Retrofitting Existing Infrastructure

Rather than building entirely new plants, municipalities are exploring modular retrofits to upgrade existing treatment plants.

- **Microplastic interceptors at sewer outflows**: Similar to river barriers, these can catch residual particles before treated water is released into natural systems.[32]
- **Sludge management**: Since microplastics accumulate in biosolids, new protocols are needed for drying, burning, or securely land-filling this material instead of spreading it on fields.[33]

C. Greywater Management

Decentralized greywater systems (for showers, sinks, laundry) are gaining attention as ways to limit plastic dispersal directly at the home or building level.

- **Filter-equipped washing machines**: Mandated in France starting in 2025, these machines trap synthetic fibers during each cycle.[34]
- **Greywater reuse**: Treated greywater used for irrigation or toilet flushing can reduce total effluent volume, lowering total plastic emissions.[35]

Strategy 5: Policy-Based Detox

Governments play a crucial role in enabling environmental detox, both by regulating pollution sources and incentivizing sustainable alternatives.

A. Bans and Restrictions

- **Single-use plastic bans**: Already enacted in many regions, these laws reduce overall plastic waste and its breakdown into microplastics.[36]
- **Micro-bead bans**: Over 100 countries have restricted the use of plastic microbeads in cosmetics and cleaning products.[37]
- **Packaging regulations**: Extended Producer Responsibility (EPR) laws require companies to take financial responsibility for their product packaging at end of life.[38]

B. Incentivizing Innovation

Governments can stimulate environmental detox by investing in research, public-private partnerships, and green technology:

- **Grants for filtration R&D**: Programs that fund universities and startups exploring novel ways to remove or neutralize plastics in the environment.[39]
- **Subsidies for biodegradable materials**: To offset the higher costs of eco-friendly alternatives and encourage wider adoption.[40]
- **Plastic offset credits**: Like carbon credits, these could allow companies to support plastic collection projects to meet reduction goals.[41]

C. Monitoring and Transparency

Environmental detox cannot succeed without robust monitoring and public accountability.

- **National plastic pollution registries**: Track sources, volumes, and types of plastic entering and exiting ecosystems.[42]

- **Citizen science networks**: Engage local volunteers to gather data on plastic density in rivers, parks, and urban centers.[43]
- **Public access dashboards**: Interactive platforms displaying real-time microplastic levels in water supplies or air quality indexes.[44]

Strategy 6: Community and NGO Initiatives

Grassroots movements are uniquely positioned to carry out localized detox efforts, mobilize volunteers, and raise public awareness.

A. Localized Plastic Audits

Many NGOs conduct plastic audits in urban and rural communities to identify high-risk areas and material types most frequently littered.

- **Brand audits**: Track which corporations contribute the most to local pollution—data that is useful for advocacy.[45]
- **Material type sorting**: Helps cities improve recycling stream separation and identify which plastics resist decomposition.[46]

B. Education and Engagement

Public education is key to long-term detox efforts:

- **Workshops**: Teach citizens how to reduce personal plastic use, install home filters, and support green policies.[47]
- **Clean-up days**: Regular events boost morale, foster environmental stewardship, and generate usable data for municipal planning.[48]
- **School programs**: Early education on the plastic lifecycle helps raise a generation of environmentally literate consumers.[49]

C. Partnerships with Industry

Some NGOs collaborate with corporations to fund and scale plastic removal projects, though these partnerships must be transparent to avoid greenwashing.

- **Sponsored river cleanups**: Companies fund initiatives in their operating areas as part of corporate social responsibility.[50]
- **Plastic-to-fuel research**: While controversial, some projects aim to convert collected plastics into usable energy in controlled, closed-loop systems.[51]

Mid-Chapter Reflections

Environmental detox is not a one-time project—it is an evolving process requiring vigilance, adaptation, and cooperation. Strategies must be tailored to the specific pollution profile of a region, whether that be marine, agricultural, urban, or industrial. Moreover, detoxification cannot stand alone; it must be part of a three-pronged approach alongside **reduction** and **replacement** of plastics.

Unlike chemical spills or oil leaks, nano and microplastics are harder to trace, easier to overlook, and more persistent. As such, detoxification must aim to be both **visible and invisible**, dealing with the plastic we can see and the microscopic debris that escapes public notice. Progress will be measured not just in kilograms of plastic removed, but in systems reformed, behaviors changed, and ecosystems stabilized.

Strategy 7: Technological Innovations in Plastic Recovery

Emerging technologies offer hope for both detecting and removing nano and microplastics at scales previously unattainable. While still in early stages or pilot deployments, these solutions provide a critical complement to policy and behavioral strategies.

A. Sensor Networks and Detection

Knowing where plastic particles concentrate is half the battle. Precision detection technologies are increasingly used to map pollution levels in real time.

- **Remote sensing**: Satellites and drones equipped with hyperspectral imaging can identify plastic waste on land and sea.[52]
- **Microplastic biosensors**: Under development in research labs, these devices aim to detect plastic particles in water using fluorescent dyes or nanosensors.[53]
- **Smart buoys and water probes**: Floating sensor systems collect samples and report data on plastic concentrations and environmental conditions like temperature or pH that affect degradation.[54]

B. Plastic-Eating Enzymes and Microorganisms

Biodegradation strategies that utilize natural or genetically engineered organisms are on the frontier of plastic detox.

- **PETase and MHETase**: Enzymes derived from *Ideonella sakaiensis* bacteria show promise in breaking down PET plastics.

Scientists are engineering more robust versions for real-world application.[55]

- **Fungal degradation**: Certain fungi, like *Aspergillus tubingensis*, can metabolize plastic in low-moisture environments, making them useful in land-based detox strategies.[56]
- **Bioreactors**: Enclosed systems that harness microbial communities to break down plastics in a controlled setting—potentially scalable for industrial waste.[57]

C. Nanomaterials for Adsorption

Advanced nanotechnology is being explored to bind or neutralize plastic pollutants in soil and water.

- **Graphene oxide**: Engineered to attract and hold microplastics through electrostatic forces.[58]
- **Magnetic nanocomposites**: These can bind with plastics and be magnetically separated from water.[59]
- **Metal-organic frameworks (MOFs)**: Porous materials with high surface areas, effective at trapping plastic monomers and additives.[60]

While these technologies are still under development or limited to research settings, they suggest a promising future for environmental detox efforts that go beyond manual or mechanical methods.

Global Case Studies

A. Norway's Wastewater Overhaul

Norway has invested in retrofitting its municipal wastewater treatment plants with high-efficiency ultrafiltration units. These systems have shown removal rates of up to 98% for microplastics, according to a 2021 report by the Norwegian Institute for Water Research.[61] The government combines this technology investment with public education and strict consumer product regulations, forming a multilayered defense against plastic pollution.

B. Kenya's Plastic Bag Ban and Urban Resilience

Since its 2017 ban on single-use plastic bags, Kenya has seen a dramatic improvement in storm drain blockage and urban cleanliness. The National Environmental Management Authority partnered with city municipalities to deploy plastic recovery teams in major urban centers, ensuring that legacy plastic waste did not undermine the progress made through legislation.[62]

C. India's Road-Building from Plastic Waste

India has constructed over 100,000 kilometers of roads using shredded plastic waste mixed with asphalt. This innovation not only reduces surface runoff of plastic particles from landfills but also enhances road durability. It's an example of circular economy thinking that links infrastructure with environmental detox.[63]

D. Indonesia's Community-Led River Booms

In Indonesia, where river pollution is a critical issue, community-built "trash booms" made of bamboo and fishing nets are used to intercept plastics in local waterways. This low-tech solution empowers villagers and has proven effective at reducing waste before it enters larger rivers or the ocean.[64]

Challenges and Limitations

While environmental detox efforts are multiplying, they face several persistent obstacles:

- **Scale mismatch**: The volume of plastic entering the environment each year far exceeds the capacity of current cleanup efforts.[65]
- **Cost**: Advanced filtration, detection, and degradation technologies are expensive and out of reach for many developing countries.[66]
- **Plastic diversity**: With thousands of polymer types, additives, and degradation products, no one-size-fits-all solution exists.[67]
- **Regulatory lag**: Technology often moves faster than the laws needed to implement or regulate its use safely.[68]
- **Greenwashing risks**: Corporations may exaggerate detox commitments to gain public approval without making real changes.[69]

Environmental detox must be embedded in a larger shift toward sustainable design, consumption, and governance. Without upstream solutions, downstream detox becomes an endless chore.

Final Thoughts

Detoxifying the environment of nano and microplastics is a monumental challenge—but not an impossible one. It will take more than just community cleanups or cutting-edge technology; it requires a coordinated, global response that spans legislation, corporate responsibility, scientific innovation, and public behavior.

Environmental detox is both reactive and proactive. It addresses legacy pollution while building systems that prevent future contamination. It demands we look inward at our lifestyles and outward at our systems. It urges us to support innovation while holding power to account.

As this chapter has outlined, there are many paths forward—some as simple as planting greenery, others as complex as engineering enzymes to eat plastic. What matters most is momentum. When individuals, cities, scientists, and governments align, detoxification becomes not just possible, but inevitable.

We are all downstream of someone else's decisions. Detoxing the environment means choosing to stop the flow—before it reaches the next generation.

Chapter 11: A Roadmap to a Plastic-Free Future

Introduction: Beyond Cleanup—Toward Transformation

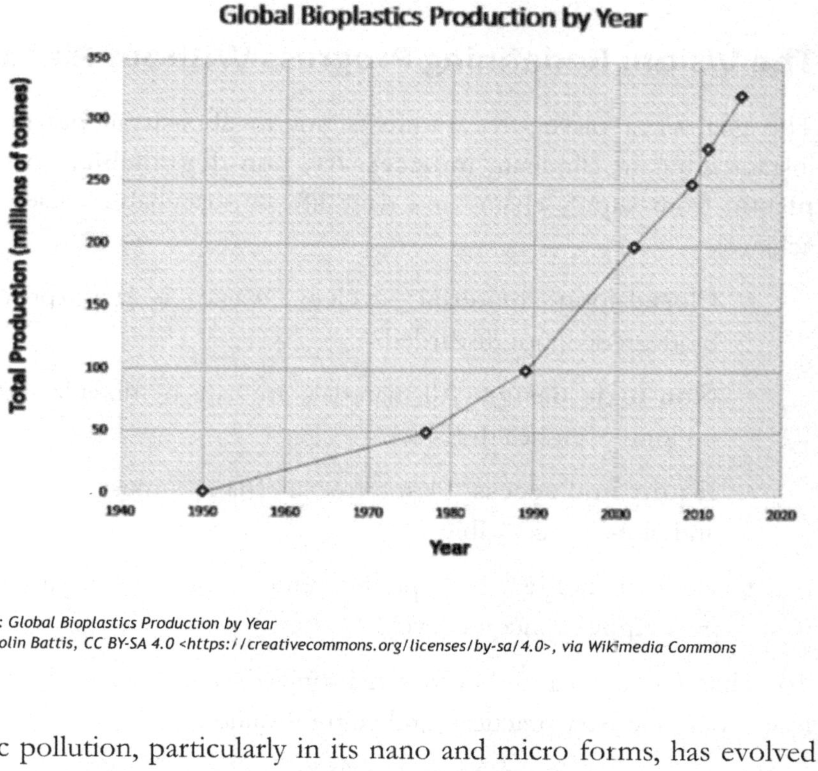

Figure 14: Global Bioplastics Production by Year
Source: Colin Battis, CC BY-SA 4.0 <https://creativecommons.org/licenses/by-sa/4.0>, via Wikimedia Commons

Plastic pollution, particularly in its nano and micro forms, has evolved into one of the most insidious environmental crises of the 21st century. While detox strategies help mitigate immediate harm, they are inherently reactive. A sustainable, long-term solution requires a shift from damage control to systemic transformation—rethinking how plastics are made, used, and disposed of at every level of society. Chapter 11 outlines this critical roadmap to a plastic-free future, focusing on global policy, technological alternatives, corporate responsibility, and collective action.

This is not merely a cleanup campaign. It is a reconfiguration of civilization's relationship with materials.

The Vision: Redefining Progress Without Plastic

The goal of a plastic-free future is not to abolish all polymer use instantly, but to eliminate **unnecessary, non-degradable, and toxic plastic** from supply chains and daily life. A sustainable society must achieve:

- **Closed-loop material cycles**: Waste is repurposed or biodegraded, not discarded.
- **Non-toxic design**: All materials are safe by default and pose no long-term health risks.
- **Equity and access**: Plastic-free alternatives must be affordable and globally accessible.
- **Zero leakage**: No plastic enters nature, regardless of geography or income level.[1]

To achieve this, societies need synchronized action across legislation, innovation, industry practices, and cultural values.[2]

Policy Pillar 1: Binding Global Treaties

One of the most promising recent developments is the effort to establish a **UN Global Treaty on Plastic Pollution**, modeled after the Paris Agreement for climate change.

A. The Treaty's Scope and Goals

- **Legally binding targets**: Reduce production and consumption of problematic plastics.
- **Monitoring mechanisms**: Standardize measure-ment and reporting across nations.
- **Support for vulnerable nations**: Include financing, technology transfer, and equitable timelines.

Negotiations launched in 2022 have gained traction, with over 170 nations expressing support. The treaty, once finalized, could become the most ambitious multilateral environmental agreement since the 1989 Montreal Protocol.[3]

B. Lessons from Other Treaties

- The **Montreal Protocol** successfully phased out ozone-depleting substances using clear deadlines and trade penalties.[4]
- The **Basel Convention** limits hazardous waste trade, and recent amendments now cover plastic waste exports.[5]

Applying similar enforcement tools and funding structures will be crucial for the success of a plastic treaty.[6]

Policy Pillar 2: National Bans and Extended Producer Responsibility (EPR)

While global treaties set the framework, national governments must pass enforceable laws to bring about localized change.

A. National Plastic Bans

Dozens of countries have implemented bans or restrictions on certain plastic items, such as:

- Single-use bags, straws, and cutlery.
- Styrofoam containers and polystyrene packaging.
- Microbeads in personal care products.

Some nations like Rwanda and Bangladesh have achieved near-complete eradication of single-use bags. Others, such as Canada and the EU bloc, have announced phased bans on multiple items.[7]

B. EPR Legislation

Extended Producer Responsibility laws make manufacturers financially and legally responsible for the lifecycle of their plastic products.

- **Deposit-return schemes**: Incentivize consumers to return plastic bottles or containers.
- **Eco-modulated fees**: Charge higher fees for harder-to-recycle materials.
- **Mandatory take-back**: Require companies to collect and recycle their own packaging.

France, Germany, and Japan are global leaders in EPR enforcement, with increasing adoption in developing economies.[8]

Policy Pillar 3: Banning Problematic Additives and Polymers

Some plastics are worse than others—due to their persistence, toxicity, or low recyclability. A successful roadmap to the future must also include **polymer prioritization.**

A. Targeting the Worst Offenders

- **PVC (Polyvinyl chloride)**: Contains toxic additives and is nearly unrecyclable.
- **Polystyrene (Styrofoam)**: Brittle, often contaminated, and not accepted in curbside recycling.
- **BPA and phthalates**: Hormone-disrupting plasticizers common in food packaging and toys.[9]

Banning or phasing out such polymers will reduce not only macro waste but also nano/microplastic generation during degradation.[10]

B. Transparency in Chemical Composition

Manufacturers are often not required to disclose all additives in plastics. Moving forward, **full material disclosure** should be mandatory:

- Ingredient labeling on packaging.
- Access to safety data sheets for recyclers.
- Certification systems for "safe plastics."

Transparency enables safer design, informed recycling, and reduced toxic exposure.[11]

Corporate Responsibility: Turning the Tide from Within

Policy-making alone cannot solve the plastic crisis without corresponding shifts in the private sector. Corporations—particularly those in consumer goods, packaging, and retail—are responsible for a large share of global plastic pollution. Yet, they also possess the resources, infrastructure, and innovation capacity to help reverse the trend. The future demands **corporate accountability paired with circular economy commitments**.[12]

A. The Plastic Footprint Revolution

Increasingly, companies are being evaluated on their **plastic footprints**, much like carbon footprints.

- **Annual reporting**: Firms disclose total plastic use, recycled content, and reduction targets.
- **Third-party verification**: NGOs such as Break Free From Plastic conduct independent plastic audits.
- **Plastic-neutral certifications**: Some companies offset plastic use by funding equivalent cleanup or recycling efforts.[13]

But critics warn that offsetting without reduction is a form of greenwashing. The most meaningful action lies in minimizing virgin plastic use at the source.[14]

B. Packaging Overhauls

Packaging represents the single largest source of plastic waste globally. Several companies have piloted packaging redesigns with measurable success.

- **Unilever** has introduced refill stations and concentrated products to reduce packaging bulk.
- **Nestlé** pledged to make all its packaging recyclable or reusable by 2025.
- **Loop** (by TerraCycle) partners with brands to offer reusable containers for delivery services.[15]

However, only a minority of companies meet their targets, and voluntary efforts are often insufficient. Governments and consumers alike must demand stronger corporate governance.[16]

C. Design for Circularity

Designing products with their end-of-life in mind is a principle of the circular economy—and a necessity in a plastic-free future.

- **Mono-material packaging**: Easier to recycle than multi-layer combinations.
- **Modular electronics**: Designed for easy disassembly and plastic part replacement.
- **Refillable and returnable systems**: Encourage long-term reuse and reduce waste.[17]

Corporate redesign isn't just environmentally necessary—it's increasingly economically viable, as raw plastic prices rise and consumer preferences shift.[18]

The Rise of Bioplastics and Alternative Materials

A plastic-free future does not mean the end of packaging, convenience, or hygienic materials. It means replacing harmful plastics with **regenerative, biodegradable, or reusable alternatives** that meet the same needs without the ecological toll.[19]

A. Bioplastics: Promise and Pitfalls

Bioplastics are plastics made from renewable resources such as corn, sugarcane, or algae. Some are biodegradable, some are not. The key distinctions include:

- **Bio-based but non-degradable**: Like bio-PET, which is chemically identical to fossil-based PET.
- **Biodegradable but not bio-based**: Such as PBAT, which may still rely on petrochemicals.
- **Compostable bioplastics**: Such as PLA, PHA, and starch blends.[20]

Challenges remain:

- **Compost confusion**: Many compostable plastics require industrial facilities and won't degrade in home compost.
- **Contamination risks**: If mixed with traditional plastics in recycling streams, they reduce recyclate quality.
- **Agricultural land use**: Large-scale bioplastic production may compete with food crops.[21]

Therefore, bioplastics are **not a silver bullet**, but one tool in a larger material transition strategy.[22]

B. Plant-Based and Mineral-Based Innovations

Beyond plastics, new materials are entering the market to meet packaging, insulation, and textile needs.

- **Mushroom packaging (mycelium)**: Biodegradable and moldable into diverse shapes.

- **Seaweed films**: Edible and water-soluble, ideal for single-serve food items.
- **Silica-coated cellulose**: Transparent, flexible, and compostable —ideal for wraps.
- **Bamboo and hemp composites**: Durable and low-impact alternatives to plastic-based construction materials.[23]

These innovations face scaling challenges, but their performance is improving rapidly, and prices are beginning to fall with broader adoption.[24]

C. Closed-Loop Materials

The circular economy doesn't always require new materials—it requires **new systems.**

- **Refill and reuse networks**: Where packaging is collected, sanitized, and reused (Loop, Algramo).
- **Deposit systems**: For durable containers, encouraging return behavior.
- **Reverse logistics**: Infrastructure that brings used materials back to the manufacturer.[25]

The challenge isn't just inventing the next material—it's redesigning supply chains to make waste obsolete.[26]

Cultural Shifts and Public Engagement

No roadmap to a plastic-free future can succeed without cultural participation. Public awareness campaigns, education reform, and activism are necessary to reshape mindsets and generate demand for alternatives.[27]

A. Education and Early Exposure

Children born today will live their entire lives under the shadow of plastic pollution. Embedding environmental literacy into school curricula is critical.

- **Plastic-free classroom policies**: Reduce exposure and model alternative behaviors.
- **Curriculum integration**: Science, civics, and economics can all explore material impacts.
- **Youth-led initiatives**: Empower students to conduct audits, cleanups, and advocacy.[28]

Several programs like **Eco-Schools** and **Plastic Free July for Schools** are scaling this model globally.[29]

B. Influencers and Media

Social media campaigns have proven powerful in shifting consumer norms.

- Hashtags like:
  #PlasticFreeJuly,
  #BreakFreeFromPlastic, and
  #ZeroWaste now reach millions.
- Viral videos of animals harmed by plastic have spurred legislation and brand boycotts.
- Eco-influencers and creators offer alternatives, reviews, and lifestyle content to support change.[30]

Cultural pressure has already influenced companies like Starbucks to eliminate plastic straws and Amazon to experiment with paper mailers.[31]

C. Civic Action and Voting Power

Citizens can push change from below through:

- **Ballot initiatives**: Local bans, packaging surcharges, or recycling mandates.
- **Consumer pressure**: Boycotts, petitions, and social accountability.
- **Electoral participation**: Supporting leaders with environmental policy commitments.[32]

Plastic pollution is not just a lifestyle issue—it is a political one. Civic engagement turns passive concern into active reform.[33]

Global Coordination and the Need for a Unified Timeline

Though local and national efforts are foundational, plastic pollution knows no borders. A bottle cap discarded in one country may wash up on another continent's shore. Microplastics in Arctic snow and Antarctic ice prove that emissions from anywhere affect everywhere. A unified timeline for phasing out harmful plastics is essential.[34]

A. Synchronizing International Goals

The **UN Environment Assembly (UNEA)** has been leading efforts to create a global plastics treaty. Such a treaty must do more than set aspirations—it must include:

- **Legally binding reduction targets.**
- **Global product standards** for recyclability and toxicity.

- **Shared accountability frameworks**, especially for multinational corporations.
- **Technical support and funding for developing nations.**[35]

Synchronization means aligning national policies with these shared objectives. Countries cannot act in isolation if plastic leakage is to be stopped within a generation.[36]

B. South-South and North-South Collaboration

High-income nations have historically exported plastic waste to lower-income countries, compounding environmental injustice. A plastic-free future must include:

- **Technology transfers** for waste management and alternative material production.
- **Funding mechanisms** such as a global plastic reduction fund.
- **Equitable phase-out timelines**, giving developing countries time and support to comply.[37]

Brazil, Indonesia, Kenya, and India have become leaders in local innovation—demonstrating that the Global South is not merely a recipient of aid, but a wellspring of scalable solutions.[38]

Timelines and Milestones: From Now to 2050

A vision is only as powerful as its roadmap. Here's a practical timeline based on current science, industrial capacity, and political will.

2026: The Foundation Phase

- Finalization and ratification of the UN global plastics treaty.
- National bans on single-use plastics in over 100 countries.

- Universal plastic footprint reporting by major corporations.
- Scale-up of compostable packaging pilot programs.[39]

2030: The Transition Phase

- 50% reduction in global virgin plastic production.
- Closure of all illegal or unmanaged plastic landfills.
- 80% of urban populations served by waste separation and recycling systems.
- Global EPR enforcement for packaging and electronics.
- Industrial-scale mycelium, seaweed, and hemp alternatives reach cost parity.[40]

2040: The Acceleration Phase

- All plastic packaging is reusable, compostable, or recyclable by design.
- International plastic waste trade is banned.
- Microplastic filters are mandatory on all washing machines and wastewater plants.
- Plastic literacy becomes a standard part of K-12 education worldwide.[41]

2050: The Regeneration Phase

- Net-zero plastic leakage into the environment.
- Global microplastic concentrations begin to decline.
- Circular economies dominate material flows.
- Ocean cleanup initiatives shift from mitigation to maintenance.

- Humans born post-2050 live in societies that no longer depend on toxic plastics.[42]

Philosophical Shift: Reclaiming Our Material Future

A plastic-free world isn't merely cleaner—it reflects a different relationship between humanity and materiality.

A. From Convenience to Consciousness

The current system prioritizes speed, cost, and convenience at the expense of planetary health. A transformed world values:

- **Durability over disposability.**
- **Repair over replacement.**
- **Design for deconstruction, not obsolescence.**[43]

Reversing our dependence on plastic requires cultural humility—an acknowledgment that modernity does not equate to sustainability.[44]

B. Ethics of Extraction and End-of-Life

Each plastic object contains the story of its origin: fossil fuels pulled from deep below the Earth, refined, extruded, and molded—only to be used for minutes and discarded for centuries.

In a plastic-free paradigm, we reclaim the ethical responsibility of **"cradle to cradle" design**. We consider where materials come from, how they serve, and where they return.[45]

Closing Reflections: A Call to Global Stewardship

The road to a plastic-free future is neither short nor easy. It demands action across every level of society—from citizens to corporations, cities to global coalitions. Yet, the convergence of public awareness, policy innovation, material science, and digital communication makes this transformation uniquely possible in our time.[46]

Imagine a world where:

- Oceans teem with life, not litter.
- Air and rain are free of synthetic fragments.
- Food packaging returns to soil without a trace.
- Children grow up never seeing plastic debris on their playgrounds.[47]

This world is not fantasy. It is a choice—one made day by day, law by law, and product by product.

The era of synthetic excess can give way to a new era of **material responsibility**, where we no longer treat Earth's ecosystems as landfills but as living partners. The roadmap is clear. The technology exists. The people are ready. All that remains is collective will.

Let this chapter serve not just as an end to a book—but as a **beginning of an era**. Your voice, your vote, your consumption, and your creativity are all pieces of the solution.

We are not plastic people by nature. We are adaptive, resilient, and capable of transformation.

Let's begin.[48]

Closing Reflections: A Call to Global Stewardship

Glossary of Terms

Chapter 1: Understanding Nano and Micro Plastics

1. Nano Plastics

Definition: Plastic particles smaller than 1 micrometer (µm), often measured in nanometers (nm).
Context: These particles originate from product fragmentation or are manufactured intentionally for industrial or medical purposes. Their small size allows them to penetrate biological membranes.

2. Micro Plastics

Definition: Plastic fragments between 1 micrometer and 5 millimeters in size.
Context: Often formed through the breakdown of larger plastic debris; found in oceans, soil, food, and water.

3. Primary Micro/Nano Plastics

Definition: Plastic particles that are **manufactured at small sizes** for specific applications (e.g., exfoliating beads, industrial abrasives).
Context: Unlike secondary plastics, these are purposefully produced.

4. Secondary Micro/Nano Plastics

Definition: Plastic particles that result from the **fragmentation of larger plastic items** through weathering, heat, or mechanical forces.
Context: Most environmental microplastics are of this type.

5. Polymer

Definition: A substance made of long, repeating chains of molecules, which are the building blocks of plastics.

Context: Common polymers include polyethylene, polypropylene, and polystyrene.

6. Monomer

Definition: The basic molecular unit that binds with others to form a polymer.
Context: Examples include ethylene and styrene, which form polyethylene and polystyrene respectively.

7. Fragmentation

Definition: The process through which larger plastic materials break down into smaller pieces.
Context: Driven by UV light, wave action, oxidation, or mechanical stress.

8. Bioavailability

Definition: The degree to which substances (like nano/micro plastics) can be absorbed by living organisms.
Context: High bioavailability in nano plastics makes them especially concerning for health.

9. Hydrophobicity

Definition: The tendency of a substance to repel water.
Context: Plastics are generally hydrophobic, which influences how they interact with biological tissues and pollutants.

10. Additives

Definition: Chemicals added to plastics during manufacturing to enhance properties like flexibility, color, or flame resistance.
Context: Common additives include phthalates and bisphenol A (BPA), which may leach out over time.

11. Sorption

Definition: The process by which plastics attract and bind chemical pollutants on their surfaces.
Context: This makes microplastics vectors for toxic substances in the environment.

12. Environmental Persistence

Definition: The resistance of a substance to natural degradation processes.
Context: Plastics persist for decades to centuries, contributing to long-term pollution.

13. Surface Area-to-Volume Ratio

Definition: A mathematical ratio expressing how much surface area a particle has relative to its volume.
Context: Smaller particles like nano plastics have a **high ratio**, increasing their reactivity and potential for chemical adsorption.

14. Biodegradation

Definition: The breakdown of materials by microorganisms.
Context: Most plastics do not biodegrade easily, but bio-based or biodegradable variants are being explored.

15. Environmental Vector

Definition: A carrier of pollutants through the environment.
Context: Microplastics can act as vectors for heavy metals, pathogens, and toxins.

Chapter 2: The Rise of Plastic Pollution

1. Bakelite

Definition: The first synthetic plastic, developed in 1907 by Leo Baekeland.
Context: Used in early electrical insulators and consumer products; marked the beginning of the synthetic polymer age.

2. Polymerization

Definition: A chemical process that combines monomers into a polymer chain.
Context: Fundamental to plastic manufacturing; responsible for creating materials like polyethylene and nylon.

3. Polyethylene (PE)

Definition: A widely used plastic polymer found in packaging, bags, and containers.
Context: Introduced in the 1930s, polyethylene is one of the most mass-produced plastics globally.

4. Polyvinyl Chloride (PVC)

Definition: A plastic polymer used in construction, pipes, and medical devices.
Context: Known for its versatility, PVC can be rigid or flexible based on its formulation.

5. Polystyrene (PS)

Definition: A lightweight plastic commonly used in disposable cutlery, containers, and foam insulation.
Context: Often found as litter and a source of microplastics due to its brittleness.

6. Plastic Boom

Definition: The post-World War II surge in plastic production and use during the 1950s and 1960s.
Context: Characterized by exponential growth in global manufacturing and consumer applications.

7. Petrochemicals

Definition: Chemical compounds derived from petroleum, used as feedstocks in plastic production.
Context: The foundation of synthetic plastic manufacturing; links plastics to fossil fuel industries.

8. Single-Use Plastics

Definition: Plastics intended for one-time use before disposal, such as straws, bags, and wrappers.
Context: Major contributors to plastic pollution; the focus of many regulatory bans and reforms.

9. Throwaway Culture

Definition: A consumer mindset characterized by the use and rapid disposal of goods, especially plastic items.
Context: Coincided with mass production and marketing in the mid-20th century.

10. Environmental Externality

Definition: A cost or effect of production or consumption that is not reflected in market prices.
Context: Plastic pollution is a classic example; the environmental damage is not paid by producers.

11. Recycling Myth

Definition: The belief that recycling alone can solve the plastic crisis.
Context: Highlighted in Chapter 2 as an inadequate solution given low recycling rates and the complexity of plastic types.

12. Great Pacific Garbage Patch

Definition: A massive accumulation of plastic debris in the North Pacific Ocean, trapped by oceanic gyres.
Context: Serves as a visual and symbolic representation of unchecked plastic waste.

13. Downcycling

Definition: The process of recycling plastic into lower-quality or less useful products.
Context: Many plastics degrade in quality after recycling and cannot be reused indefinitely.

14. Plastic Footprint

Definition: The total amount of plastic waste an individual, organization, or country generates.
Context: Analogous to a carbon footprint, it is used to assess and reduce plastic-related impacts.

15. Linear Economy

Definition: An economic model based on "take, make, dispose" with little reuse or recycling.
Context: Contrasted with a circular economy in sustainability discussions.

Chapter 3: Where Plastics Lurk: A Global Ubiquity

1. Ubiquity

Definition: The state of being present everywhere.
Context: Refers to the global pervasiveness of nano and microplastics in all environments—from oceans to mountaintops.

2. Atmospheric Deposition

Definition: The process by which airborne particles settle onto land or water surfaces.
Context: A major route through which microplastics contaminate remote regions, including mountains and polar areas.

3. Indoor Dust

Definition: Fine particles suspended in the air or settled in indoor environments.
Context: Often contains synthetic fibers and microplastics from carpets, furniture, and clothing.

4. Synthetic Fibers

Definition: Man-made fibers such as polyester, nylon, and acrylic.
Context: Shed from clothing and textiles during wear or washing; a major source of microplastics in water and air.

5. Wastewater Treatment Plants (WWTPs)

Definition: Facilities that clean sewage and industrial effluent before discharge into the environment.
Context: Remove a significant portion of microplastics but do not capture all; remaining particles enter aquatic systems.

6. Sludge Application

Definition: The use of treated sewage sludge as fertilizer on agricultural land.
Context: A pathway for microplastics to enter soils, especially when biosolids contain synthetic residues.

7. Trophic Transfer

Definition: The movement of substances such as plastics through food chains.
Context: Microplastics ingested by smaller organisms are passed on to predators, concentrating at higher levels.

8. Plastic Inhalation

Definition: The intake of airborne microplastics through breathing.
Context: Documented in urban settings and indoor environments, particularly where synthetic textiles are abundant.

9. Food Contamination

Definition: The presence of unwanted substances in food products.
Context: Microplastics have been found in salt, honey, sugar, seafood, and bottled water.

10. Bottled Water Contamination

Definition: The presence of microplastics in commercial bottled water.
Context: Studies have found microfibers and fragments in nearly all tested brands, regardless of region.

11. Marine Snow

Definition: Organic particles and debris that sink from the upper ocean to the seafloor.
Context: Can carry microplastics to deep ocean environments, contributing to contamination of benthic habitats.

12. Sentinel Species

Definition: Organisms used to detect environmental hazards due to their sensitivity to pollutants.
Context: Mussels, fish, and seabirds are often used to monitor microplastic exposure in marine ecosystems.

13. Biofilm Formation

Definition: The colonization of microplastic surfaces by microorganisms.
Context: Changes the plastic's buoyancy and ecological interactions, potentially making it more bioavailable to marine life.

14. Freshwater Contamination

Definition: The pollution of rivers, lakes, and streams with plastic particles.
Context: Often overlooked in early research, freshwater systems are now recognized as major conduits for plastic transport.

15. Household Sources

Definition: Domestic activities and products contributing to microplastic release.
Context: Includes laundering clothes, using personal care products, vacuuming carpets, and food packaging disposal.

Chapter 4: Sources of Contamination

1. Fragmentation

Definition: The physical breakdown of larger plastic items into smaller pieces.
Context: A primary mechanism by which plastic waste becomes micro- and nanoplastics in the environment.

2. Weathering

Definition: The degradation of materials due to exposure to environmental elements like sunlight, wind, water, and temperature changes.
Context: UV radiation from sunlight contributes significantly to the fragmentation of plastics.

3. Tire Wear Particles

Definition: Microscopic rubber and plastic particles shed from vehicle tires during use.
Context: A major source of microplastic pollution, particularly in urban runoff.

4. Synthetic Textiles

Definition: Man-made fabrics such as polyester, nylon, and acrylic.
Context: Shed fibers during washing, contributing to microplastic contamination in wastewater systems.

5. Microbeads

Definition: Tiny plastic beads formerly used in cosmetics and personal care products (e.g., face scrubs, toothpaste).
Context: Now banned in many countries, but still present in legacy products and certain regions.

6. Industrial Pellets (Nurdles)

Definition: Small plastic pellets used as raw material in the manufacturing of plastic products.
Context: Commonly lost during transport or processing, leading to environmental contamination.

7. Packaging Waste

Definition: Discarded plastic materials used for wrapping, protecting, or containing goods.
Context: Includes bags, wrappers, bottles, and containers—among the most common sources of plastic pollution.

8. Fishing Gear Debris

Definition: Abandoned, lost, or discarded fishing equipment such as nets, lines, and traps.
Context: Degrades into microplastics over time and threatens marine wildlife through entanglement or ingestion.

9. Construction Debris

Definition: Waste material from construction sites, including plastic tarps, insulation, and coatings.
Context: Often underreported but significant in contributing plastic fragments to soil and water.

10. Urban Runoff

Definition: Rainwater or melted snow that flows over urban surfaces, collecting pollutants including microplastics.
Context: A major conduit for tire particles, synthetic fibers, and packaging debris into waterways.

11. Agricultural Plastics

Definition: Plastic products used in farming, such as mulch films, irrigation tubes, and greenhouses.
Context: Breakdown of these materials contributes to soil microplastic contamination.

12. Personal Care Products

Definition: Everyday items like shampoos, body washes, and exfoliants that may contain synthetic polymers.
Context: Although microbeads are banned in many places, soluble and insoluble synthetic additives are still common.

13. Road Marking Paint

Definition: Paints containing plastic resins used to stripe roads and highways.
Context: Wears off over time and contributes microplastic particles to nearby soil and water.

14. Industrial Abrasives

Definition: Plastic particles used in processes like sandblasting or polishing.
Context: Can be released into the environment if not properly filtered or contained.

15. Paint Flakes

Definition: Microscopic or small plastic-rich chips shed from painted surfaces.
Context: Found in marine environments, often originating from ships, buildings, or infrastructure.

Chapter 5: Pathways into the Human Body

1. Ingestion

Definition: The process of consuming substances through the mouth.
Context: One of the primary pathways through which nano and microplastics enter the human body via food, beverages, and utensils.

2. Inhalation

Definition: The act of breathing in air containing particles or gases.
Context: Airborne microplastics and synthetic fibers present in household and urban dust can be inhaled into the lungs.

3. Dermal Absorption

Definition: The uptake of substances through the skin.
Context: A debated but possible route for very small nanoplastics, especially in contact with cosmetic or medical products.

4. Nanoplastics

Definition: Plastic particles less than 1 micrometer (1000 nanometers) in size.
Context: Due to their size, nanoplastics can cross biological membranes and interact with cells and tissues.

5. Translocation

Definition: The movement of particles from one part of the body to another.
Context: Nanoplastics can translocate from the digestive tract to the bloodstream or lymphatic system.

6. Bloodstream Penetration

Definition: The ability of a substance to enter the circulatory system.
Context: Once in the bloodstream, nanoplastics may reach organs such as the liver, spleen, or brain.

7. Blood–Brain Barrier (BBB)

Definition: A selective semipermeable membrane that protects the brain from toxins and pathogens.
Context: Some studies suggest nanoplastics might breach the BBB due to their size and chemical properties.

8. Placental Transfer

Definition: The movement of substances from a pregnant person to the fetus through the placenta.
Context: Microplastics have been detected in placental tissue, raising concerns about prenatal exposure.

9. Mucosal Barriers

Definition: Protective linings found in the respiratory and gastrointestinal tracts.
Context: These barriers can be compromised by plastic particles, leading to increased vulnerability to pathogens and toxins.

10. Endocytosis

Definition: A process by which cells engulf substances from their surroundings.
Context: A possible mechanism by which nanoplastics enter human cells.

11. Microfold (M) Cells

Definition: Specialized cells in the intestine that sample and transport particles to immune cells.

Context: A suspected route for nanoplastic absorption into the lymphatic system.

12. Gastrointestinal Tract (GI Tract)

Definition: The digestive system path from mouth to anus.
Context: A primary exposure route for ingested plastics, with potential for local and systemic effects.

13. Bioavailability

Definition: The extent to which a substance can be absorbed and used by the body.
Context: Determines how much of an ingested or inhaled plastic particle can affect internal systems.

14. Adsorption

Definition: The accumulation of substances, such as toxins or heavy metals, on the surface of microplastics.
Context: Increases the potential for plastics to deliver harmful chemicals to the body upon ingestion or inhalation.

15. Medical Exposure

Definition: Contact with plastics through medical procedures or devices.
Context: Includes plastic-based implants, tubing, and drug delivery systems that may shed microscopic particles.

Chapter 6: Impact on Human Health

1. Toxicology

Definition: The study of the adverse effects of chemical substances on living organisms.
Context: Used to assess the risks associated with micro- and nanoplastics and their chemical additives.

2. Endocrine Disruptors

Definition: Chemicals that interfere with hormone systems.
Context: Common plastic additives like BPA and phthalates can disrupt reproductive, metabolic, and developmental functions.

3. Immune Dysregulation

Definition: Abnormal functioning of the immune system, either overactive or suppressed.
Context: Exposure to plastic particles may trigger chronic inflammation or impair immune responses.

4. Inflammation

Definition: The body's immune response to harmful stimuli.
Context: Microplastics have been shown to cause inflammation in tissues, particularly in the lungs and intestines.

5. Oxidative Stress

Definition: A condition where the body has an imbalance between free radicals and antioxidants.
Context: Nanoplastics can promote oxidative stress, which damages cells, proteins, and DNA.

6. Neurotoxicity

Definition: Harmful effects on the nervous system.
Context: Emerging studies suggest nanoplastics may affect brain function and development.

7. Cytotoxicity

Definition: The quality of being toxic to cells.
Context: Lab tests show plastic particles can impair cell membranes, reduce viability, and alter cell cycles.

8. Genotoxicity

Definition: The ability of a substance to damage genetic information in cells.
Context: Some plastic-associated chemicals have demonstrated genotoxic effects in in-vitro studies.

9. Bioaccumulation

Definition: The gradual accumulation of substances, such as plastics or chemicals, in an organism.
Context: Repeated exposure to microplastics may lead to buildup in organs over time.

10. Chronic Exposure

Definition: Continuous or repeated contact with a substance over a long period.
Context: More likely to cause long-term health effects than acute (short-term) exposure.

11. Particle Translocation

Definition: The movement of particles from one part of the body to another.

Context: Nanoplastics may cross barriers like the gut lining or blood-brain barrier to affect distant organs.

12. Allergic Sensitization

Definition: The process by which the immune system becomes overly responsive to a normally harmless substance.
Context: Plastic particles or additives may exacerbate allergies or trigger new immune sensitivities.

13. Microbiome Disruption

Definition: Alteration of the natural microbial community in the human body, especially in the gut.
Context: Microplastics may shift the balance of beneficial and harmful bacteria, impacting digestion and immunity.

14. Tumor Promotion

Definition: The enhancement of cellular environments that favor the growth of cancerous cells.
Context: Some additives in plastics are suspected to have carcinogenic or tumor-promoting effects.

15. Exposure Threshold

Definition: The minimum quantity of a substance required to elicit a detectable biological response.
Context: Ongoing research aims to identify the exposure thresholds at which plastics begin to affect human health.

Chapter 7: Ecological and Biodiversity Effects

1. Biodiversity

Definition: The variety of living species within a given ecosystem, including animals, plants, fungi, and microorganisms.
Context: Micro- and nanoplastics threaten biodiversity by impacting multiple species and trophic levels.

2. Ecosystem Services

Definition: The benefits that humans derive from healthy ecosystems (e.g., pollination, water purification, soil fertility).
Context: Disruption by microplastics can diminish these services, affecting food production and environmental stability.

3. Trophic Transfer

Definition: The movement of substances like nutrients or contaminants through different levels of a food chain.
Context: Microplastics can be ingested by small organisms and transferred up the food web to predators.

4. Bioaccumulation

Definition: The buildup of substances, such as plastic particles or toxins, in an organism over time.
Context: Observed in aquatic life and soil organisms exposed to persistent plastic pollution.

5. Soil Microbiota

Definition: Microorganisms living in the soil, including bacteria, fungi, and archaea.
Context: Affected by plastic presence, which can alter nutrient cycling and soil health.

6. Earthworms

Definition: Soil-dwelling invertebrates that aerate soil and aid in decomposition.
Context: Shown to ingest microplastics, resulting in reduced reproduction and burrowing activity.

7. Fibrous Plastics

Definition: Plastic fragments that are elongated, often originating from synthetic textiles or ropes.
Context: These can interfere with soil structure and animal digestion when ingested.

8. Detritivores

Definition: Organisms that feed on decomposing organic material.
Context: Species like earthworms and some insects are exposed to plastics through their feeding habits.

9. Microalgae

Definition: Microscopic algae that serve as primary producers in aquatic environments.
Context: Their photosynthesis can be impaired by plastic-induced turbidity and competition with biofilms.

10. Zooplankton

Definition: Small aquatic animals that drift with currents and serve as a vital food source for larger marine species.
Context: Prone to ingesting microplastics, which reduce their energy intake and reproductive output.

11. Biofilms

Definition: Communities of microorganisms that grow on surfaces, including plastics.
Context: Change microbial interactions and make plastics more attractive to aquatic feeders.

12. Endocrine Disruption (in Wildlife)

Definition: Hormonal interference in animals caused by environmental pollutants.
Context: Associated with plastic additives and absorbed toxins in microplastics.

13. Feeding Efficiency

Definition: A measure of how effectively an organism can obtain and process food.
Context: Reduced in animals that ingest plastics, leading to energy deficits and developmental issues.

14. Aquatic Vertebrates

Definition: Fish, amphibians, and other animals with backbones living in water.
Context: Experience immune suppression and reproductive harm due to plastic exposure.

15. Ecosystem Fragmentation

Definition: The breakdown of ecological networks or habitats into isolated patches.
Context: May be accelerated by cumulative plastic pollution altering species distributions and survival.

Chapter 8: Hotspots and Safe Havens

1. Hotspot

Definition: A geographic area with unusually high concentrations of a particular element or activity—in this case, micro- and nanoplastics.
Context: Coastal cities, major rivers, and industrial zones are often plastic pollution hotspots.

2. Safe Haven

Definition: A location with minimal contamination or environmental disruption.
Context: Remote mountain ranges, polar regions, or protected marine reserves may act as temporary refuges from plastic pollution.

3. Plastic Footprint

Definition: The total amount of plastic waste generated by an individual, region, or system.
Context: Used to compare and analyze the contribution of different regions to global plastic contamination.

4. Hydrological Transport

Definition: The movement of substances through water systems such as rivers, lakes, and oceans.
Context: Plastics travel long distances via rainfall runoff, stormwater, and currents.

5. Atmospheric Deposition

Definition: The process by which airborne particles settle onto land or water.
Context: Microplastics have been found to travel hundreds of miles through wind and settle far from their origin.

6. Urban Runoff

Definition: Rainwater or melted snow that flows over urban surfaces and into water systems, carrying pollutants.
Context: A major pathway for microplastics from tire wear, construction debris, and synthetic fibers.

7. Coastal Sediments

Definition: Layers of particulate matter found near shorelines.
Context: Often contain accumulated plastic particles from nearby land and sea-based sources.

8. Wastewater Treatment Plants (WWTPs)

Definition: Facilities that process sewage and industrial effluents before releasing water back into the environment.
Context: A key point of plastic interception, though many small particles escape final filtration.

9. Sediment Core Sampling

Definition: A method for studying layers of deposited material in water bodies to understand historical pollution.
Context: Used to track the rise of microplastic concentrations over decades.

10. Trophic Exposure

Definition: The transfer of pollutants like plastics through consumption across different levels of the food chain.
Context: Measured in both aquatic and terrestrial environments to assess ecological risk.

11. Polar Amplification

Definition: The phenomenon where polar regions warm and respond more intensely to climate change.
Context: It can influence plastic fragmentation and transport via melting glaciers and shifting currents.

12. Ocean Gyres

Definition: Large systems of circulating ocean currents that trap floating debris.
Context: Examples include the Great Pacific Garbage Patch, which harbors massive quantities of fragmented plastics.

13. Remote Sensing

Definition: The use of satellite or aerial technology to monitor environmental changes and pollution patterns.
Context: Increasingly used to model and track microplastic distribution globally.

14. Watershed

Definition: A geographic area where all rainfall drains to a common outlet such as a river, lake, or bay.
Context: Plays a major role in the concentration and direction of plastic pollution.

15. Environmental Baseline

Definition: A reference point for assessing current pollution levels against historic or pre-industrial conditions.
Context: Helps identify new hotspots and evaluate the success of mitigation efforts.

Chapter 9: Strategies for Personal Detox

1. Personal Detox

Definition: A set of practices intended to reduce the body's burden of environmental toxins.
Context: Includes minimizing plastic exposure and adopting health-supportive behaviors.

2. Endocrine Disruptors

Definition: Chemicals that interfere with hormonal systems, often found in plastics.
Context: Avoided by reducing use of plastics labeled with BPA or phthalates.

3. BPA-Free

Definition: Products that do not contain bisphenol A, a known endocrine-disrupting chemical.
Context: Seen as safer alternatives for food containers and water bottles.

4. Activated Charcoal

Definition: A highly porous substance used to trap toxins in the digestive tract.
Context: Sometimes used as a supplement during detox protocols to aid elimination of ingested toxins.

5. Bioaccumulation

Definition: The gradual buildup of chemicals or particles in an organism over time.
Context: Plastic-related chemicals may bioaccumulate in organs and fatty tissues.

6. Chelation

Definition: A medical or nutritional process of binding and removing heavy metals or toxins from the body.
Context: Some detox regimens aim to chelate plastic-related contaminants, although clinical efficacy remains under study.

7. Antioxidants

Definition: Compounds that neutralize harmful free radicals in the body.
Context: Found in detox-supportive foods like berries, green tea, and leafy greens.

8. Detox Pathways

Definition: The body's mechanisms for eliminating waste, such as through the liver, kidneys, skin, and lungs.
Context: Supporting these pathways helps reduce accumulated toxins including plastic residues.

9. Intermittent Fasting

Definition: A dietary pattern that alternates between periods of eating and fasting.
Context: Some claim it enhances cellular repair and toxin elimination, although evidence is emerging.

10. Sauna Therapy

Definition: The use of heat, typically through dry or infrared saunas, to promote sweating and detoxification.
Context: Suggested to aid in the removal of fat-stored toxins, though plastic-specific effects are not fully studied.

11. Gut Microbiome

Definition: The collection of microorganisms in the digestive tract that influence immunity, digestion, and detoxification.
Context: A healthy gut can help reduce the reabsorption of toxins like plastic additives.

12. Nutraceuticals

Definition: Products derived from food sources with extra health benefits in addition to their basic nutritional value.
Context: May include supplements marketed for detox or liver support.

13. Plastic-Free Lifestyle

Definition: A way of living that intentionally avoids the use of plastic products.
Context: Core to preventing personal exposure to micro- and nanoplastics.

14. Clean Label

Definition: A term used in food and consumer products to signify limited, recognizable, and natural ingredients.
Context: Often includes avoiding plasticizers and artificial preservatives.

15. Environmental Exposure

Definition: Contact with pollutants through air, water, food, or skin.
Context: A central concept in understanding how plastics enter and affect the body.

Chapter 10: Strategies for Environmental Detox

1. Environmental Detox

Definition: The systematic removal or reduction of harmful pollutants —including plastics—from ecosystems.
Context: Involves cleanup technologies, community initiatives, and policy enforcement.

2. Bioremediation

Definition: The use of microorganisms or plants to break down pollutants in soil, water, or air.
Context: Considered a promising strategy for degrading certain plastic compounds naturally.

3. Plastic-Eating Enzymes

Definition: Engineered or naturally occurring proteins that can degrade plastic polymers into harmless byproducts.
Context: Used in experimental cleanup approaches, especially for PET (polyethylene terephthalate).

4. Microplastic Filtration

Definition: The physical removal of tiny plastic particles from air or water using mesh screens, membrane filters, or electrostatic systems.
Context: Applied in wastewater treatment plants and stormwater systems.

5. Zero Waste

Definition: A philosophy that encourages the redesign of resource life cycles so that all products are reused or recycled.
Context: Reduces the generation of plastics at the source.

6. Circular Economy

Definition: An economic model focused on designing out waste, keeping products in use, and regenerating natural systems.
Context: Involves extended producer responsibility and incentivizes plastic reuse.

7. Extended Producer Responsibility (EPR)

Definition: A policy approach requiring manufacturers to take responsibility for the entire lifecycle of their products, including disposal.
Context: Often mandates plastic recovery or recycling targets.

8. Beach Cleanup Initiatives

Definition: Organized efforts—often volunteer-based—to remove plastics and other waste from coastal environments.
Context: Help reduce marine plastic input and raise public awareness.

9. Green Infrastructure

Definition: Nature-based systems such as wetlands, bioswales, or rain gardens that treat runoff and filter pollutants.
Context: Reduces the flow of microplastics into natural water systems.

10. Industrial Effluent Management

Definition: Monitoring and treatment of liquid waste released by factories and manufacturing plants.
Context: Essential to prevent plastic particles and chemicals from entering water bodies.

11. Plastic Capture Devices

Definition: Equipment like trash booms, river skimmers, or floating barriers designed to intercept plastics before they reach open waters.
Context: Used in rivers, harbors, and wastewater outlets.

12. Smart Waste Sorting

Definition: The use of AI or sensor-driven systems to automatically identify and sort recyclable and non-recyclable materials.
Context: Increases plastic recycling rates and decreases contamination.

13. Compostable Alternatives

Definition: Materials designed to decompose under natural or industrial composting conditions, replacing conventional plastics.
Context: Used in packaging, utensils, and agricultural films.

14. Environmental Policy Enforcement

Definition: The implementation and policing of laws that restrict plastic use or mandate cleanup actions.
Context: Varies by country or municipality and includes plastic bag bans or micro-bead restrictions.

15. Public-Private Partnerships (PPPs)

Definition: Collaborations between government bodies and private companies to achieve environmental goals.
Context: Often fund or operate large-scale plastic waste collection and treatment systems.

Chapter 11: A Roadmap to a Plastic-Free Future

1. Plastic-Free Economy

Definition: An economic system in which plastic use is minimized or eliminated through sustainable alternatives, regulations, and public behavior.
Context: The goal of global sustainability movements and long-term environmental planning.

2. International Treaty on Plastic Pollution

Definition: A proposed or enacted multinational agreement aimed at controlling and reducing plastic waste and its impacts.
Context: Modeled on treaties like the Paris Agreement, this aims to unify plastic management goals globally.

3. Corporate Responsibility

Definition: The ethical obligation of companies to manage the environmental and social impact of their products, including plastics.
Context: Often emphasized through voluntary commitments or regulated by EPR laws.

4. Innovation Ecosystem

Definition: A network of entrepreneurs, researchers, companies, and governments collaborating to solve environmental challenges.
Context: Critical to developing plastic alternatives, recycling technology, and global solutions.

5. Green Chemistry

Definition: The design of chemical products and processes that reduce or eliminate hazardous substances.
Context: Supports the creation of non-toxic, biodegradable plastic alternatives.

6. Bioplastics

Definition: Plastics made from renewable biological sources (e.g., corn starch, sugarcane) rather than petroleum.
Context: Promoted as eco-friendly, though not always biodegradable or compostable.

7. Policy Levers

Definition: Regulatory tools used by governments to influence market behavior, such as taxes, subsidies, bans, and standards.
Context: Includes plastic bag bans, import restrictions, or incentives for sustainable packaging.

8. Public Engagement

Definition: The active involvement of communities and individuals in shaping policies and behaviors related to plastic use.
Context: Drives momentum through education, activism, and consumer demand shifts.

9. Lifecycle Assessment (LCA)

Definition: A method of evaluating the environmental impacts of a product from raw material to disposal.
Context: Used to determine the true sustainability of alternatives to plastic.

10. Greenwashing

Definition: Marketing that falsely portrays a product or practice as environmentally friendly.
Context: A risk when companies adopt superficial changes without addressing core sustainability issues.

11. Plastic Offsetting

Definition: Similar to carbon offsetting, this involves funding plastic collection or recycling efforts to counterbalance plastic use.
Context: Controversial if used to delay real systemic change.

12. Stakeholder Alignment

Definition: The coordination of diverse parties (governments, corporations, NGOs, citizens) around shared environmental goals.
Context: Essential for large-scale success in plastic reduction.

13. Regenerative Design

Definition: A philosophy that aims not just to sustain but to restore and renew ecosystems through product and system design.
Context: Goes beyond "less harm" to "net positive" environmental impact.

14. Sustainable Procurement

Definition: The practice of purchasing goods and services that minimize environmental impact across their life cycle.
Context: Includes choosing vendors with low-plastic or circular business models.

15. Just Transition

Definition: A framework ensuring that the move toward sustainability includes fair outcomes for workers, communities, and economies.
Context: Important for avoiding unintended social consequences as industries shift away from plastic.

Master Glossary: Nano and Micro Plastics Book

A comprehensive, alphabetized glossary.

Activated Charcoal
A highly porous substance used to trap toxins in the digestive tract, sometimes used during detox protocols.

Antioxidants
Compounds that neutralize harmful free radicals in the body, often found in detox-supportive foods.

Beach Cleanup Initiatives
Organized efforts, often volunteer-based, to remove plastics and waste from coastal areas.

Bioaccumulation
The gradual buildup of chemicals or particles in an organism over time.

Biodegradable Plastics
Plastics designed to decompose under natural environmental conditions via microbial action.

Bioplastics
Plastics made from renewable biological sources like corn starch or sugarcane.

Bioremediation
The use of living organisms, like microbes or plants, to detoxify polluted environments.

BPA-Free
Products that do not contain bisphenol A, a chemical with endocrine-disrupting properties.

Circular Economy
An economic system aimed at minimizing waste and making the most of resources through reuse and recycling.

Clean Label
Indicates food or product ingredients that are simple, recognizable, and free from harmful additives.

Compostable Alternatives
Materials that can safely decompose in composting environments, replacing plastics.

Corporate Responsibility
The ethical obligation of businesses to mitigate environmental harm from their products.

Detox Pathways
The body's mechanisms for eliminating waste through the liver, kidneys, lungs, and skin.

Earthworms and Detritivores
Organisms that consume and recycle decomposing organic material, affected by soil plastics.

Ecotoxicology
The study of toxic effects of substances, including plastics, on the natural environment.

Endocrine Disruptors
Chemicals that interfere with the body's hormonal systems.

Environmental Detox
The systematic removal of pollutants like microplastics from ecosystems.

Environmental Exposure
Contact with pollutants through air, water, food, or skin.

Environmental Policy Enforcement
Implementation of regulations designed to control or reduce pollution.

Extended Producer Responsibility (EPR)
A policy requiring manufacturers to take responsibility for their products throughout their lifecycle.

Green Chemistry
The design of chemical products and processes that reduce or eliminate hazardous substances.

Green Infrastructure
Nature-based solutions, such as wetlands, that help filter and manage pollutants.

Greenwashing
Misleading marketing to make products appear more environmentally friendly than they are.

Gut Microbiome
The diverse community of microorganisms in the human digestive tract, crucial for health and detox.

Innovation Ecosystem
Collaborative environment where stakeholders develop and deploy sustainable technologies.

Intermittent Fasting
A dietary practice alternating between periods of eating and fasting, sometimes used in detox.

International Treaty on Plastic Pollution
A proposed global agreement to limit and manage plastic production and waste.

Just Transition
A concept ensuring fairness and equity for workers and communities during environmental shifts.

Lifecycle Assessment (LCA)
Method for evaluating the environmental impact of a product across its entire life cycle.

Microplastic Filtration
The removal of tiny plastic particles from air or water using physical or electrostatic filters.

Microplastics
Plastic particles between 1 micrometer and 5 millimeters in size, resulting from product breakdown.

Nano Plastics
Plastic particles smaller than 1 micrometer, often measured in nanometers.

Nutraceuticals
Food-based products with added health benefits beyond basic nutrition.

Personal Detox
Individual efforts to reduce the body's toxin load through lifestyle and dietary changes.

Plastic Capture Devices
Equipment like booms or skimmers used to intercept plastic waste in waterways.

Plastic-Free Economy
A system in which plastics are replaced by sustainable alternatives in all sectors.

Plastic Offsetting
Funding efforts to remove or recycle plastic to counterbalance one's own plastic use.

Plastisphere
Microbial communities that colonize plastic surfaces in natural environments.

Policy Levers
Regulatory mechanisms, such as bans or taxes, used to influence plastic use and production.

Primary Microplastics
Small plastic particles intentionally manufactured for use in products like cosmetics.

Public Engagement
Involvement of the general population in environmental decision-making and actions.

Public-Private Partnerships (PPPs)
Collaborative projects between governments and businesses to address environmental challenges.

Regenerative Design
Design philosophy that aims to restore and renew ecosystems, going beyond sustainability.

Sauna Therapy
Use of heat to promote sweating and detoxification through the skin.

Smart Waste Sorting
Technologies using AI or sensors to automatically separate recyclable from non-recyclable waste.

Soil Microbial Communities
Bacteria, fungi, and other microorganisms in soil ecosystems, affected by microplastic presence.

Stakeholder Alignment
Cooperation among various groups—governments, corporations, and citizens—to achieve sustainability goals.

Sustainable Procurement
Choosing suppliers and products that meet environmental and ethical standards.

Trophic Transfer
Movement of contaminants through levels of the food chain as predators consume contaminated prey.

Wastewater Treatment Plants
Facilities that remove contaminants—including microplastics—from sewage before it enters natural water bodies.

Zero Waste
A lifestyle and policy approach aiming to send no waste to landfills or incinerators through reuse and recycling.

Biography

Robert Sudar is an author and advocate focused on the growing crisis of global plastic pollution. Driven by the alarming lack of public awareness regarding nano and micro plastics, he compiled years of global data to create a roadmap for individual and to advance systemic change. **Robert** lives in the United States and is committed to a future where the air, water, and soil are once again pristine, and the food supply is free of plastic contaminants.

Chapter References

Chapter 1 References:

1. Wright, S. L., & Kelly, F. J. (2017). Plastic and Human Health: A Micro Issue? *Environmental Science & Technology*, 51(12), 6634–6647.

2. Andrady, A. L. (2011). Microplastics in the marine environment. *Marine Pollution Bulletin*, 62(8), 1596–1605.

3. Napper, I. E., & Thompson, R. C. (2016). Release of synthetic microplastic plastic fibres from domestic washing machines: Effects of fabric type and washing conditions. *Marine Pollution Bulletin*, 112(1–2), 39–45.

4. Cole, M. et al. (2011). Microplastics as contaminants in the marine environment: A review. *Marine Pollution Bulletin*, 62(12), 2588–2597.

5. Gigault, J. et al. (2018). Current opinion: What is a nanoplastic? *Environmental Pollution*, 235, 1030–1034.

6. Hartmann, N. B. et al. (2019). Are We Speaking the Same Language? Recommendations for a Definition and Categorization Framework for Plastic Debris. *Environmental Science & Technology*, 53(3), 1039–1047.

7. Lusher, A. L. et al. (2013). Occurrence of microplastics in the gastrointestinal tract of pelagic and demersal fish from the English Channel. *Marine Pollution Bulletin*, 67(1–2), 94–99.

8. Mattsson, K. et al. (2017). Brain damage and behavioural disorders in fish induced by plastic nanoparticles delivered through the food chain. *Scientific Reports*, 7, 11452.

9. Rochman, C. M. et al. (2013). Ingested plastic transfers hazardous chemicals to fish and induces hepatic stress. *Scientific Reports*, 3, 3263.

10. Koelmans, A. A. et al. (2016). Microplastic as a Vector for Chemicals in the Aquatic Environment: Critical Review and Model-Supported Reinterpretation of Empirical Studies. *Environmental Science & Technology*, 50(7), 3315–3326.

11. Geyer, R., Jambeck, J. R., & Law, K. L. (2017). Production, use, and fate of all plastics ever made. *Science Advances*, 3(7), e1700782.

12. Hermabessiere, L. et al. (2017). Occurrence and effects of plastic additives on marine environments and organisms: A review. *Chemosphere*, 182, 781–793.

13. Carney Almroth, B. M. et al. (2018). Quantifying shedding of synthetic fibers from textiles; a source of microplastics released into the environment. *Environmental Science and Pollution Research*, 25, 1191–1199.

14. Dris, R. et al. (2016). Synthetic fibers in atmospheric fallout: A source of microplastics in the environment? *Marine Pollution Bulletin*, 104(1–2), 290–293.

15. Tokiwa, Y. et al. (2009). Biodegradability of plastics. *International Journal of Molecular Sciences*, 10(9), 3722–3742.

16. Barnes, D. K. A. et al. (2009). Accumulation and fragmentation of plastic debris in global environments. *Philosophical Transactions of the Royal Society B*, 364(1526), 1985–1998.

17. Hidalgo-Ruz, V. et al. (2012). Microplastics in the marine environment: A review of the methods used for identification and quantification. *Environmental Science & Technology*, 46(6), 3060–3075.

18. Käppler, A. et al. (2016). Identification of microplastics by FTIR and Raman microscopy: A novel approach. *Journal of Spectroscopy*, 2016, 1–10.

19. Enders, K. et al. (2020). Mass spectrometry for the analysis of microplastic particles and their associated chemicals. *TrAC Trends in Analytical Chemistry*, 124, 115806.

20. Fries, E. et al. (2013). Identification of polymer types and additives in marine microplastic particles using pyrolysis-GC/MS and scanning electron microscopy. *Environmental Science: Processes & Impacts*, 15(10), 1949–1956.

21. Mintenig, S. M. et al. (2018). Low numbers of microplastics detected in drinking water from ground water sources. *Science of the Total Environment*, 648, 631–635.

22. Prata, J. C. (2018). Airborne microplastics: Consequences to human health? *Environmental Pollution*, 234, 115–126.

23. Eriksen, M. et al. (2014). Plastic pollution in the world's oceans: More than 5 trillion plastic pieces weighing over 250,000 tons afloat at sea. *PLOS ONE*, 9(12), e111913.

24. Woodall, L. C. et al. (2014). The deep sea is a major sink for microplastic debris. *Royal Society Open Science*, 1(4), 140317.

25. Cole, M. et al. (2013). Microplastic ingestion by zooplankton. *Environmental Science & Technology*, 47(12), 6646–6655.

26. Lusher, A. L. et al. (2015). Microplastics in Arctic polar waters: The first reported values of particles in surface and sub-surface samples. *Scientific Reports*, 5, 14947.

27. Velzeboer, I. et al. (2014). Nanoplastics in the aquatic environment. Critical review. *Current Opinion in Environmental Science & Health*, 1, 17–23.

28. Nizzetto, L. et al. (2016). Are agricultural soils dumps for microplastics of urban origin? *Environmental Science & Technology*, 50(20), 10777–10779.

29. Zubris, K. A. V., & Richards, B. K. (2005). Synthetic fibers as an indicator of land application of sludge. *Environmental Pollution*, 138(2), 201–211.

30. Huerta Lwanga, E. et al. (2016). Microplastics in the terrestrial ecosystem: A hidden threat. *Scientific Reports*, 6, 33834.

31. Rillig, M. C. et al. (2017). Microplastic pollution in terrestrial ecosystems. *Science*, 356(6335), 1430–1431.

32. de Souza Machado, A. A. et al. (2019). Microplastics can change soil properties and affect plant performance. *Environmental Science & Technology*, 53(10), 6044–6052.

33. Lehmann, A. et al. (2020). Microplastic effects on plants. *New Phytologist*, 228(1), 48–53.

34. Dris, R. et al. (2015). A first overview of textile fibers, including microplastics, in indoor and outdoor environments. *Environmental Pollution*, 200, 654–662.

35. Kole, P. J. et al. (2017). Wear and tear of tyres: A stealthy source of microplastics in the environment. *International Journal of Environmental Research and Public Health*, 14(10), 1265.

36. Allen, S. et al. (2019). Atmospheric transport and deposition of microplastics in a remote mountain catchment. *Nature Geoscience*, 12(5), 339–344.

37. Bergmann, M. et al. (2019). White and wonderful? Microplastics prevail in snow from the Alps to the Arctic. *Science Advances*, 5(8), eaax1157.

38. Gasperi, J. et al. (2018). Microplastics in air: Are we breathing it in? *Current Opinion in Environmental Science & Health*, 1, 1–5.

39. Provencher, J. F. et al. (2018). Marine birds and plastic debris in Canada: A national synthesis and a way forward. *Environmental Reviews*, 26(4), 361–370.

40. Wright, S. L. et al. (2013). The physical impacts of microplastics on marine organisms: A review. *Environmental Pollution*, 178, 483–492.

41. Watts, A. J. R. et al. (2016). Uptake and retention of microplastics by the shore crab *Carcinus maenas*. *Environmental Science & Technology*, 48(15), 8823–8830.

42. Cox, K. D. et al. (2019). Human consumption of microplastics. *Environmental Science & Technology*, 53(12), 7068–7074.

43. Smith, M. et al. (2018). Microplastics in seafood and the implications for human health. *Current Environmental Health Reports*, 5(3), 375–386.

44. Teuten, E. L. et al. (2009). Transport and release of chemicals from plastics to the environment and to wildlife. *Philosophical Transactions of the Royal Society B*, 364(1526), 2027–2045.

45. Rochester, J. R. (2013). Bisphenol A and human health: A review of the literature. *Reproductive Toxicology*, 42, 132–155.

46. Holmes, L. A. et al. (2012). Adsorption of trace metals to plastic resin pellets in the marine environment. *Environmental Pollution*, 160(1), 42–48.

47. Bhattacharya, P. et al. (2010). Physical adsorption of charged plastic nanoparticles affects algal photosynthesis. *Journal of Physical Chemistry C*, 114(39), 16556–16561.

48. Forte, M. et al. (2016). Nanoplastics exposure causes mitochondrial dysfunction in fish hepatocytes. *Toxicology in Vitro*, 35, 56–62.

49. Schirinzi, G. F. et al. (2017). Cytotoxic effects of nanoplastics and microplastics on human cell lines. *Environmental Research*, 159, 579–587.

50. Hwang, J. et al. (2019). Potential toxicity of polystyrene microplastic particles. *Scientific Reports*, 9, 7391.

51. Pitt, J. A. et al. (2018). Maternal transfer of nanoplastics to offspring in zebrafish. *Environmental Science & Technology*, 52(18), 11078–11086.

52. Revel, M. et al. (2018). Micro(nano)plastics: A threat to human health? *Current Opinion in Environmental Science & Health*, 1, 17–23.

53. European Chemicals Agency (ECHA). (2020). Annex XV Restriction Report – Intentionally added microplastics.

54. United States Congress. (2015). *Microbead-Free Waters Act of 2015*, Public Law No: 114-114.

55. Government of Canada. (2020). Science assessment of plastic pollution.

56. Koelmans, A. A. et al. (2019). Guidance for the harmonized sampling and analysis of microplastics. *Water Research*, 155, 410–422.

57. Leslie, H. A. (2014). Review of microplastics in cosmetics. *IVM Institute for Environmental Studies Report*.

58. Sharma, S., & Chatterjee, S. (2017). Microplastic pollution, a threat to marine ecosystem and human health: A short review. *Environmental Science and Pollution Research*, 24(27), 21530–21547.

59. BBC. (2017). *Blue Planet II*.

60. Craig, R. K. (2018). A comparative guide to the emerging law of plastic pollution. *Ecology Law Quarterly*, 45(1), 127–180.

61. Break Free From Plastic. (2021). Annual Brand Audit Report.

62. UNESCO. (2020). Ocean literacy for all: A toolkit.

63. Araujo, C. F. et al. (2018). Identification of microplastics using Raman spectroscopy: Latest developments and future prospects. *Water Research*, 142, 426–440.

64. Munno, K. et al. (2020). Increasing the resolution and scope of microplastic analyses using near-infrared spectroscopy. *Journal of Hazardous Materials*, 393, 122433.

65. Primpke, S. et al. (2018). Reference database design for automated analysis of microplastic samples based on FTIR spectroscopy. *Analytical and Bioanalytical Chemistry*, 410(21), 5131–5141.

66. La Nasa, J. et al. (2021). Pyrolysis-GC-MS for the analysis of microplastics. *TrAC Trends in Analytical Chemistry*, 133, 116049.

67. Fries, E., & Dekiff, J. H. (2016). Determination of microplastics using thermogravimetric analysis. *Analytical Methods*, 8(29), 6346–6350.

68. Mason, S. A. et al. (2016). Nanoplastic particle detection using nanoparticle tracking analysis. *Marine Pollution Bulletin*, 104(1–2), 103–110.

69. Gigault, J. et al. (2016). Current opinion: What is a nanoplastic? *Environmental Pollution*, 215, 273–275.

70. Labib, S. et al. (2020). Adaptation of flow cytometry for detection of polystyrene nanoplastics in biological matrices. *Environmental Toxicology and Chemistry*, 39(12), 2487–2493.

71. Mitrano, D. M. et al. (2019). Emerging analytical methods to detect and quantify nanoplastics. *Current Opinion in Environmental Science & Health*, 1, 36–41.

72. Holmes, L. A. et al. (2014). Adsorption of trace metals to plastic resin pellets in the marine environment. *Environmental Pollution*, 185, 155–160.

73. Teuten, E. L. et al. (2009). Transport and release of chemicals from plastics to the environment and to wildlife. *Philosophical Transactions of the Royal Society B*, 364(1526), 2027–2045.

74. Zettler, E. R. et al. (2013). "Plastisphere": Microbial communities on plastic marine debris. *Environmental Science & Technology*, 47(13), 7137–7146.

75. Lithner, D. et al. (2011). Environmental and health hazard ranking and assessment of plastic polymers based on chemical composition. *Science of the Total Environment*, 409(18), 3309–3324.

76. Ellen MacArthur Foundation. (2016). *The New Plastics Economy: Rethinking the future of plastics*.

77. Niaounakis, M. (2013). *Biopolymers: Applications and trends*. William Andrew Publishing.

78. Sharma, C. et al. (2020). Edible packaging materials: Sustainable alternatives for plastic. *Food Packaging and Shelf Life*, 24, 100477.

79. Bocken, N. et al. (2016). Product design and business model strategies for a circular economy. *Journal of Industrial and Production Engineering*, 33(5), 308–320.

80. OECD. (2021). *Extended Producer Responsibility: Updated Guidance for Efficient Waste Management*.

81. Van Ewijk, S. & Stegemann, J. A. (2016). Limitations of the waste hierarchy for achieving absolute reductions in material throughput. *Journal of Cleaner Production*, 132, 122–128.

82. Watkins, E. et al. (2017). Effectiveness of policy measures to reduce microplastic pollution. *European Commission Report*.

83. Ragaert, K. et al. (2017). Mechanical and chemical recycling of solid plastic waste. *Waste Management*, 69, 24–58.

84. Rahimi, A. & García, J. M. (2017). Chemical recycling of waste plastics for new materials production. *Nature Reviews Chemistry*, 1(6), 0046.

85. Gardiner, S. M. (2006). A perfect moral storm: Climate change, intergenerational ethics and the problem of moral corruption. *Environmental Values*, 15(3), 397–413.

86. Liboiron, M. (2018). Waste is not 'matter out of place'. *Discard Studies Blog*.

87. Jamieson, D. (2008). The moral and political challenges of climate change. *NYU Environmental Law Journal*, 16, 439–469.

88. Geyer, R. et al. (2017). Production, use, and fate of all plastics ever made. *Science Advances*, 3(7), e1700782.

89. Shrivastava, P. & Ivanova, M. (2015). Flourishing: A vision for sustainability. *Organization & Environment*, 28(2), 123–129.

90. Thompson, R. C. et al. (2009). Our plastic age. *Philosophical Transactions of the Royal Society B*, 364(1526), 1973–1976.

91. Lambert, S. et al. (2017). The environmental fate of microplastics: A mechanistic perspective. *Environmental Science & Technology*, 51(12), 6620–6621.

92. Jambeck, J. R. et al. (2015). Plastic waste inputs from land into the ocean. *Science*, 347(6223), 768–771.

93. Rochman, C. M. (2018). Microplastics research—from sink to source. *Science*, 360(6384), 28–29.

94.

Chapter 2 References:

1. Freinkel, S. (2011). *Plastic: A Toxic Love Story*. Houghton Mifflin Harcourt.

2. Geyer, R. et al. (2017). Production, use, and fate of all plastics ever made. *Science Advances*, 3(7), e1700782.

3. Thompson, R. C. et al. (2009). Our plastic age. *Philosophical Transactions of the Royal Society B*, 364(1526), 1973–1976.

4. Jambeck, J. R. et al. (2015). Plastic waste inputs from land into the ocean. *Science*, 347(6223), 768–771.

5. Meikle, J. L. (1995). *American Plastic: A Cultural History*. Rutgers University Press.

6. Fenichell, S. (1996). *Plastic: The Making of a Synthetic Century*. HarperBusiness.

7. Kauffman, G. B. (2006). Leo Baekeland and the birth of Bakelite. *Chemistry International*, 28(2), 4–6.

8. Andrady, A. L. & Neal, M. A. (2009). Applications and societal benefits of plastics. *Philosophical Transactions of the Royal Society B*, 364(1526), 1977–1984.

9. Freinkel, S. (2011). *Plastic: A Toxic Love Story.*

10. DuPont Archives. (2020). History of Nylon.

11. Brydson, J. A. (1999). *Plastics Materials*. Butterworth-Heinemann.

12. Andrady, A. L. (2003). *Plastics and the Environment*. John Wiley & Sons.

13. Geyer, R. (2020). *The Business of Less: The Role of Companies and Households on a Planet in Peril*. Routledge.

14. Strasser, S. (1989). *Satisfaction Guaranteed: The Making of the American Mass Market*. Smithsonian Institution.

15. Hopewell, J. et al. (2009). Plastics recycling: Challenges and opportunities. *Philosophical Transactions of the Royal Society B*, 364(1526), 2115–2126.

16. PlasticsEurope. (2020). *Plastics – the Facts 2020.*

17. Cohen, L. (2003). *A Consumers' Republic: The Politics of Mass Consumption in Postwar America*. Vintage.

18. Rhodes, C. J. (2018). Plastic pollution and potential solutions. *Science Progress*, 101(3), 207–260.

19. Thompson, R. C. et al. (2004). Lost at sea: Where is all the plastic? *Science*, 304(5672), 838.

20. Jamieson, A. J. et al. (2019). Microplastics and human health. *Marine Pollution Bulletin*, 138, 1–2.

21. Carpenter, E. J. & Smith, K. L. (1972). Plastics on the Sargasso Sea surface. *Science*, 175(4027), 1240–1241.

22. Ibid.

23. Laist, D. W. (1987). Overview of the biological effects of lost and discarded plastic debris in the marine environment. *Marine Pollution Bulletin*, 18(6), 319–326.

24. Keep America Beautiful. (1971). *Crying Indian PSA Campaign*.

25. Ibid.

26. Browne, M. A. et al. (2011). Accumulation of microplastic on shorelines worldwide: sources and sinks. *Environmental Science & Technology*, 45(21), 9175–9179.

27. Andrady, A. L. (2011). Microplastics in the marine environment. *Marine Pollution Bulletin*, 62(8), 1596–1605.

28. Karami, A. et al. (2017). Microplastic contamination in table salt from different countries. *Scientific Reports*, 7, 46173.

29. Schwabl, P. et al. (2018). Assessment of microplastic concentrations in human stool. *United European Gastroenterology Journal*, 6(8), 1124–1130.

30. Cox, K. D. et al. (2019). Human consumption of microplastics. *Environmental Science & Technology*, 53(12), 7068–7074.

31. Prata, J. C. (2018). *Airborne microplastics: Consequences to human health?* Environmental Pollution, 234, 115–126.\

32. Wright, S. L., & Kelly, F. J. (2017). *Plastic and human health: A micro issue?* Environmental Science & Technology, 51(12), 6634–6647.

33. Koelmans, A. A., et al. (2015). *Nanoplastics in the aquatic environment.* Critical Reviews in Environmental Science and Technology, 45(22), 2037–2079.

34. Yong, C. Q. Y., et al. (2020). *Toxicity of microplastics and nanoplastics in mammalian systems.* International Journal of Environmental Research and Public Health, 17(5), 1509.

35. Geyer, R., et al. (2017). *Production, use, and fate of all plastics ever made.* Science Advances, 3(7), e1700782.

36. OECD (2022). *Global plastics outlook: Economic drivers, environmental impacts and policy options.*

37. UNEP (2023). *Turning off the tap: How the world can end plastic pollution and create a circular economy.*

38. Pew Charitable Trusts & SYSTEMIQ (2020). *Breaking the plastic wave: A comprehensive assessment of pathways towards stopping ocean plastic pollution.*

39. Jambeck, J. R., et al. (2015). *Plastic waste inputs from land into the ocean.* Science, 347(6223), 768–771.

40. Hopewell, J., et al. (2009). *Plastics recycling: Challenges and opportunities.* Philosophical Transactions of the Royal Society B, 364(1526), 2115–2126.

41. Brooks, A. L., et al. (2018). *The Chinese import ban and its impact on global plastic waste trade.* Science Advances, 4(6), eaat0131.

42. Ibid.

43. Lebreton, L. C., et al. (2018). *Evidence that the Great Pacific Garbage Patch is rapidly accumulating plastic.* Scientific Reports, 8(1), 4666.

44. Eriksen, M., et al. (2014). *Plastic pollution in the world's oceans.* PLOS ONE, 9(12), e111913.

45. van Sebille, E., et al. (2015). *A global inventory of small floating plastic debris.* Environmental Research Letters, 10(12), 124006.

46. Allen, S., et al. (2020). *Atmospheric transport and deposition of microplastics in a remote mountain catchment.* Nature Geoscience, 13, 339–344.

47. Ibid.

48. Schuyler, Q. A., et al. (2014). Risk analysis of plastic ingestion by sea turtles. *Conservation Biology.*

49. Wilcox, C., et al. (2015). Threat of plastic pollution to seabirds is global, pervasive, and increasing. *PNAS.*

50. Jacobsen, J. K., et al. (2010). Observations of entangled gray whales. *Marine Pollution Bulletin.*

51. Rochman, C. M., et al. (2014). Endocrine disrupting chemicals in plastic. *Environmental Science & Technology.*

52. Zhou, Q., et al. (2020). Microplastics in agricultural soils. *Science of the Total Environment*.

53. Wright, S. L., & Kelly, F. J. (2017). Plastic and human health: A micro issue? *Environmental Science & Technology*.

54. Ragusa, A., et al. (2021). Plastic particles in human placenta. *Environment International*.

55. Talsness, C. E., et al. (2009). Components of plastic: Human health concerns. *Philosophical Transactions B*.

56. European Commission. (2017). *Scientific Advice on Plastics and Health*.

57. UNEP. (2014). Valuing plastics: The business case for measuring, managing and disclosing plastic use.

58. Beaumont, N. J., et al. (2019). Global ecological, social and economic impacts of marine plastic. *Marine Pollution Bulletin*.

59. Center for International Environmental Law. (2019). *Plastic and climate: The hidden costs*.

60. UNEP. (2021). *Bioplastics and the circular economy*.

61. Hopewell, D., Dvorak, R., & Kosior, E. (2009). Plastics recycling: challenges and opportunities. *Philosophical Transactions of the Royal Society B*, 364(1526), 2115-2126.

62. Geyer, R., Jambeck, J. R., & Law, K. L. (2017). Production, use, and fate of all plastics ever made. *Science Advances*, 3(7), e1700782.

63. American Chemistry Council. (2021). Understanding Plastic Resin Codes.

64. Brooks, A. L., Wang, S., & Jambeck, J. R. (2018). The Chinese import ban and its impact on global plastic waste trade. *Science Advances*, 4(6), eaat0131.

65. Greenpeace. (2019). Trashed: Southeast Asia's imported plastic crisis.

66. European Commission. (2019). Directive (EU) 2019/904 on the reduction of the impact of certain plastic products on the environment.

67. Government of Canada. (2022). Single-use Plastics Prohibition Regulations.

68. UNEP. (2018). Kenya's Plastic Bag Ban: Enforcement and Lessons Learned.

69. Oregon Department of Environmental Quality. (2021). Extended Producer Responsibility for Packaging.
70. UNEP. (2022). Global Plastic Pollution Treaty Resolution.
71. Netflix. (2016–2021). *A Plastic Ocean, The Story of Plastic, Seaspiracy.*
72. Break Free From Plastic. (2023). Movement Building and Social Change.
73. EarthDay.org. (2023). Youth and the Plastic Crisis.
74. Center for International Environmental Law. (2020). Plastic and Climate: The Hidden Costs.
75. Tournier, V. et al. (2020). An engineered PET depolymerase to break down and recycle plastic bottles. *Nature*, 580(7802), 216–219.
76. Ellis, L. D. et al. (2021). Chemical and biological catalysis for plastics recycling and upcycling. *Nature Catalysis*, 4, 539–556.
77. IBM Research. (2021). Blockchain for Sustainable Supply Chains.
78. TerraCycle. (2023). Loop: Revolutionizing Packaging.
79. National Geographic. (2021). Education and Citizen Science to Combat Plastic Pollution.
80. World Economic Forum. (2016). The New Plastics Economy: Rethinking the Future of Plastics.
81. Ellen MacArthur Foundation. (2016). The New Plastics Economy: Rethinking the Future of Plastics.
82. Freinkel, S. (2011). *Plastic: A Toxic Love Story.* Houghton Mifflin Harcourt.

Chapter 3 References:

1. Bergmann, M., et al. (2019). "Plastic pollution in the ocean." *Annual Review of Marine Science.*
2. Jamieson, A.J., et al. (2017). "Plastic contamination of deep-sea amphipods." *Royal Society Open Science.*
3. Lebreton, L., et al. (2018). "Evidence that the Great Pacific Garbage Patch is rapidly accumulating plastic." *Scientific Reports.*

4. Chae, Y., & An, Y.J. (2018). "Effects of micro- and nanoplastics on aquatic ecosystems." *Water Research*.

5. Mason, S.A., et al. (2018). "Synthetic polymer contamination in bottled water." *Frontiers in Chemistry*.

6. Lechner, A., & Ramler, D. (2015). "The Danube so white—A geophysical estimate of plastic pollution in the Danube River." *Environmental Pollution*.

7. Dris, R., et al. (2016). "A first overview of textile fibers, including microplastics, in indoor and outdoor environments." *Environmental Pollution*.

8. Allen, S., et al. (2019). "Atmospheric transport and deposition of microplastics in a remote mountain catchment." *Nature Geoscience*.

9. Prata, J.C. (2018). "Airborne microplastics: Consequences to human health?" *Environmental Pollution*.

10. Wright, S.L., & Kelly, F.J. (2017). "Plastic and human health: A micro issue?" *Environmental Science & Technology*.

11. Nizzetto, L., et al. (2016). "Are agricultural soils dumps for microplastics?" *Environmental Science & Technology*.

12. Zhang, G.S., & Liu, Y.F. (2018). "The distribution of microplastics in soil aggregate fractions in southwestern China." *Science of the Total Environment*.

13. Van Cauwenberghe, L., & Janssen, C.R. (2014). "Microplastics in bivalves cultured for human consumption." *Environmental Pollution*.

14. Karami, A., et al. (2017). "Microplastics in table salt from different countries." *Scientific Reports*.

15. Liebezeit, G., & Liebezeit, E. (2013). "Non-pollen particulates in honey and sugar." *Food Additives & Contaminants: Part A*.

16. Li, L., et al. (2020). "Nanoplastics in agricultural soils: Sources, interactions, and impacts." *Science of the Total Environment*.

17. Talsness, C.E., et al. (2009). "Components of plastic: Human health concerns." *Philosophical Transactions of the Royal Society B*.

18. Leslie, H.A., et al. (2016). "Microplastics in cosmetics: Environmental hazards and regulatory gaps." *Environmental International*.

19. Galloway, T.S., & Lewis, C.N. (2016). "Marine microplastics spell big problems for future generations." *PNAS.*

20. Napper, I.E., & Thompson, R.C. (2016). "Release of synthetic microplastic fibres from domestic washing machines." *Marine Pollution Bulletin.*

21. Kapp, K.J., & Miller, R.Z. (2020). "Elevated microfiber release from polyester fabrics in tumble dryers." *Environmental Pollution.*

22. Boucher, J., & Friot, D. (2017). "Primary microplastics in the oceans: A global evaluation of sources." *IUCN Report.*

23. Prata, J.C., et al. (2020). "Sources of microplastics in indoor environments." *Science of the Total Environment.*

24. Wright, S.L., & Kelly, F.J. (2017). "Plastic and human health: A micro issue?" *Environmental Science & Technology.*

25. Gasperi, J., et al. (2018). "Assessment of atmospheric microplastic deposition in urban environments." *Nature Communications.*

26. Ragusa, A., et al. (2021). "Plasticenta: First evidence of microplastics in human placenta." *Environment International.*

27. Leslie, H.A., et al. (2022). "Discovery and quantification of plastic particle pollution in human blood." *Environment International.*

28. Sharma, S., & Chatterjee, S. (2017). "Microplastic pollution, a threat to marine ecosystem and human health: A short review." *Environmental Science and Pollution Research.*

29. Brahney, J., et al. (2020). "Plastic rain in protected areas of the United States." *Science.*

30. Barboza, L.G.A., et al. (2018). "Marine microplastic debris: An emerging issue for food security, food safety and human health." *Marine Pollution Bulletin.*

31. Bergmann, M., et al. (2019). "High quantities of microplastic in Arctic deep-sea sediments from the HAUSGARTEN observatory." *Environmental Science & Technology.*

32. Allen, S., et al. (2019). "Atmospheric transport and deposition of microplastics in a remote mountain catchment." *Nature Geoscience.*

33. Dris, R., et al. (2016). "Synthetic fibers in atmospheric fallout: A source of microplastics in the environment?" *Marine Pollution Bulletin*.

34. Liu, F., et al. (2021). "Release of microplastics from plastic waste recycling processes." *Journal of Hazardous Materials*.

35. Pivokonsky, M., et al. (2018). "Occurrence of microplastics in raw and treated drinking water." *Science of the Total Environment*.

36. Zheng, Y., et al. (2022). "Nanoplastics generated from municipal waste incineration: Characteristics and potential risks." *Environmental Pollution*.

37. European Commission. (2018). "A European Strategy for Plastics in a Circular Economy."

38. Haider, T.P., et al. (2019). "Plastics of the future? The impact of biodegradable polymers on the environment and on society." *Angewandte Chemie International Edition*.

39. Kalčíková, G., et al. (2017). "Wastewater treatment plant effluents as source of cosmetic polyethylene microbeads to freshwater." *Chemosphere*.

40. Abbasi, S., et al. (2019). "Microplastics in household dust: Sources, pathways, and human exposure." *Environmental Science & Technology Letters*.

41. Fadare, O.O., et al. (2020). "Microparticles and microplastics released from daily-use plastic food containers." *Science of the Total Environment*.

42. Cowger, W., et al. (2020). "Microplastic monitoring: Challenges, progress, and future directions." *Environmental Science & Technology*.

43. UNEP. (2021). "From Pollution to Solution: A Global Assessment of Marine Litter and Plastic Pollution."

44. The Plastic Soup Foundation. (2023). "Citizen science projects tackling plastic pollution." *plasticsoupfoundation.org*.

Chapter 4 References:

1. Geyer, R., et al. (2017). "Production, use, and fate of all plastics ever made." *Science Advances*.

2. Andrady, A. L. (2011). "Microplastics in the marine environment." *Marine Pollution Bulletin*.

3. OECD. (2022). "Global Plastics Outlook: Economic Drivers, Environmental Impacts and Policy Options."

4. Jambeck, J. R., et al. (2015). "Plastic waste inputs from land into the ocean." *Science*.

5. European Commission. (2018). "A European Strategy for Plastics in a Circular Economy."

6. Napper, I. E., & Thompson, R. C. (2016). "Release of synthetic microplastic fibers from domestic washing machines." *Marine Pollution Bulletin*.

7. Carr, S. A., et al. (2016). "Transport of microplastic particles in wastewater treatment plants." *Water Research*.

8. Henry, B., et al. (2019). "Microfiber pollution and the textile industry." *Textile World*.

9. Kole, P. J., et al. (2017). "Wear and tear of tyres: A stealthy source of microplastics in the environment." *International Journal of Environmental Research and Public Health*.

10. Wagner, S., et al. (2018). "Tire wear particles in the aquatic environment." *Science of the Total Environment*.

11. Evangeliou, N., et al. (2020). "Atmospheric transport is a major pathway of microplastics to remote regions." *Nature Communications*.

12. Rochman, C. M., et al. (2019). "The ecological impacts of microplastics on marine life." *Journal of Hazardous Materials*.

13. Karlsson, T. M., et al. (2018). "Nurdle pollution on coastlines: global scale and pathways." *Marine Pollution Bulletin*.

14. Lambert, S., & Wagner, M. (2016). "Characterisation of nanoplastics during the degradation of polystyrene." *Chemosphere*.

15. Prata, J. C. (2018). "Microplastics in wastewater: from sources to treatment." *Science of the Total Environment*.

16. Vianello, A., et al. (2019). "Simulating microplastic exposure in humans." *Environmental Science & Technology*.

17. Corradini, F., et al. (2019). "Evidence of microplastic accumulation in agricultural soil from plastic mulch." *Science of the Total Environment*.

18. Steinmetz, Z., et al. (2016). "Plastic mulching in agriculture." *Environmental Science & Technology.*

19. Zhang, G. S., & Liu, Y. F. (2018). "The distribution of microplastics in soil." *Environmental Pollution.*

20. Liu, M., et al. (2020). "Controlled release of pesticides using microplastic carriers." *Pesticide Biochemistry and Physiology.*

21. Murphy, F., et al. (2016). "Wastewater treatment works as a source of microplastics in the environment." *Environmental Science & Technology.*

22. UNEP. (2015). "Plastic in cosmetics: Are we polluting the environment through our personal care?"

23. Leslie, H. A. (2014). "Review of microplastics in cosmetics." *IVM Institute for Environmental Studies.*

24. Mintenig, S. M., et al. (2017). "Microplastics in toothpaste and exfoliants." *Water Research.*

25. Gago, J., et al. (2018). "Anthropogenic particles in cleaning products." *Chemosphere.*

26. Enfrin, M., et al. (2019). "Release and breakdown of microplastics from personal care products." *Environmental Science: Water Research & Technology.*

27. Gewert, B., et al. (2015). "Weathering of plastic marine debris and its impact on particle generation." *Environmental Science & Technology.*

28. Andrady, A. L. (2011). "Microplastics in the marine environment." *Marine Pollution Bulletin.*

29. Fok, L., & Cheung, P. K. (2015). "Degradation of plastic debris in river shorelines." *Water Research.*

30. Tosin, M., et al. (2012). "Biodegradation of plastics in controlled environments." *Polymer Degradation and Stability.*

31. Sun, J., et al. (2021). "Landfill leachate as a significant source of microplastics." *Science of the Total Environment.*

32. Rillig, M. C., & Lehmann, A. (2020). "Microplastic in terrestrial ecosystems." *Science.*

33. Allen, S., et al. (2019). "Atmospheric transport and deposition of microplastics in a remote mountain catchment." *Nature Geoscience*.

34. Prata, J. C., et al. (2020). "Indoor microplastic pollution: presence, sources, and health effects." *Environmental International*.

35. Zhang, Y., et al. (2021). "Redistribution of microplastics through indoor air systems." *Building and Environment*.

36. Bergmann, M., et al. (2019). "White and wonderful? Microplastics prevail in snow from the Alps to the Arctic." *Science Advances*.

37. Brahney, J., et al. (2020). "Plastic rain in protected areas of the United States." *Science*.

38. UNEP. (2016). "Marine plastic debris and microplastics: Global lessons and research to inspire action and guide policy change."

39. Macfadyen, G., et al. (2009). "Abandoned, lost or otherwise discarded fishing gear." *FAO Fisheries and Aquaculture Technical Paper*.

40. Ytreberg, E., et al. (2020). "Antifouling paints as a source of microplastics." *Marine Pollution Bulletin*.

41. Wang, W., et al. (2018). "Marine discharge of microplastics from ship-related cleaning activities." *Environmental Science & Technology Letters*.

42. Enders, K., et al. (2015). "Transfer of microplastics across ocean regions by ballast water." *Marine Pollution Bulletin*.

43. Veiga, J. M., et al. (2016). "Tourism and recreational activities as a source of marine litter." *Ocean & Coastal Management*.

44. Browne, M. A., et al. (2011). "Accumulation of microplastic on shorelines." *Environmental Science & Technology*.

45. Rech, S., et al. (2015). "Microplastic debris in beach sediments: effects of seasonal tourist activity." *Environmental Pollution*.

46. Ambrosini, R., et al. (2019). "Microplastic contamination in snow from Western Italian Alps." *Environmental Science & Technology*.

47. Silva-Cavalcanti, J. S., et al. (2017). "Beach management and microplastic pollution." *Marine Pollution Bulletin*.

48. Geyer, R., et al. (2017). "Production, use, and fate of all plastics ever made." *Science Advances*.

49. Sun, J., et al. (2019). "Microplastics in wastewater treatment plants: Detection, occurrence and removal." *Water Research*.

50. Singh, N., et al. (2017). "Recycling of plastic solid waste: A state of art review and future applications." *Composites Part B: Engineering*.

51. Brooks, A. L., et al. (2018). "The Chinese import ban and its impact on global plastic waste trade." *Science Advances*.

52. Henry, B., et al. (2019). "Microfiber pollution and the fashion industry." *Environmental Health Perspectives*.

53. Vianello, A., et al. (2021). "Plastic combustion: A neglected source of environmental nanoplastics." *Environmental Pollution*.

54. Hale, R. C., et al. (2002). "Persistent pollutants in plastic combustion residues." *Environmental Science & Technology*.

55. Kroll, J. H., et al. (2020). "Air quality and plastic combustion." *Journal of Geophysical Research: Atmospheres*.

56. Li, J., et al. (2020). "Indoor microplastics after fire events." *Environmental Science: Processes & Impacts*.

57. Wiedinmyer, C., et al. (2014). "Global emissions from open burning of domestic waste." *Environmental Science & Technology*.

58. Horton, A. A., et al. (2017). "Microplastics in freshwater and terrestrial environments." *Environmental Science & Technology*.

59. Lambert, S., & Wagner, M. (2016). "Characterisation of nanoplastics during the degradation of polystyrene." *Chemosphere*.

60. Yong, C. Q. Y., et al. (2020). "Toxicity of micro- and nanoplastics in mammalian systems." *International Journal of Environmental Research and Public Health*.

61. Hou, L., et al. (2021). "Nanoplastics in the environment Toxic effects and mechanisms." *Environmental Science: Nano*.

Chapter 5 References:

1. Cox, K.D. et al. (2019). Human Consumption of Microplastics. *Environmental Science & Technology*, 53(12), 7068–7074.

2. Pivokonsky, M. et al. (2018). Occurrence of microplastics in raw and treated drinking water. *Science of the Total Environment*, 643, 1644–1651.

3. Mason, S.A. et al. (2018). Synthetic Polymer Contamination in Bottled Water. *Frontiers in Chemistry*, 6, 407.

4. Mintenig, S.M. et al. (2019). Microplastics in drinking water: A review. *Water Research*, 155, 410–422.

5. Van Cauwenberghe, L. & Janssen, C.R. (2014). Microplastics in bivalves cultured for human consumption. *Environmental Pollution*, 193, 65–70.

6. Barboza, L.G.A. et al. (2020). Microplastics in wild fish from North East Atlantic Ocean. *Marine Pollution Bulletin*, 154, 111072.

7. Karami, A. et al. (2017). Microplastic and heavy metals in table salt from Chinese markets. *Scientific Reports*, 7, 46173.

8. Liebezeit, G. & Liebezeit, E. (2014). Synthetic particles as contaminants in German beers. *Food Additives & Contaminants: Part A*, 31(9), 1574–1578.

9. Hahladakis, J.N. & Iacovidou, E. (2018). An overview of chemical additives present in plastics: Migration, release, fate and environmental impact. *Journal of Hazardous Materials*, 344, 179–199.

10. Li, J. et al. (2021). Microplastics in Takeaway Food Containers. *Environmental Science & Technology*, 55(2), 1045–1055.

11. EFSA Panel on Contaminants in the Food Chain (2016). Presence of microplastics and nanoplastics in food. *EFSA Journal*, 14(6), e04501.

12. Prata, J.C. (2018). Airborne microplastics: Consequences to human health? *Environmental Pollution*, 234, 115–126.

13. Wright, S.L. & Kelly, F.J. (2017). Plastic and Human Health: A Micro Issue? *Environmental Science & Technology*, 51(12), 6634–6647.

14. Vianello, A. et al. (2019). Simulating human exposure to indoor airborne microplastics using a breathing thermal manikin. *Scientific Reports*, 9, 8670.

15. Zhang, J. et al. (2020). Infant exposure to microplastics via hand-to-mouth behavior. *Environmental Science & Technology*, 54(7), 4043–4050.

16. Evangeliou, N. et al. (2020). Atmospheric transport is a major pathway of microplastics to remote regions. *Nature Communications*, 11, 3381.

17. Allen, S. et al. (2019). Atmospheric transport of microplastics and implications for human exposure. *Nature Geoscience*, 12, 339–344.

18. Napper, I.E. et al. (2015). Characterisation, quantity and sorptive properties of microplastics extracted from cosmetics. *Marine Pollution Bulletin*, 99(1–2), 178–185.

19. Leslie, H.A. (2014). Review of microplastics in cosmetics. *RIVM Report 2014-0157.*

20. GESAMP (2015). Sources, fate and effects of microplastics in the marine environment: A global assessment. *IMO/FAO/UNESCO-IOC/UNIDO/WMO/IAEA/UN/UNEP Joint Group of Experts on the Scientific Aspects of Marine Environmental Protection Reports and Studies*, No. 90.

21. Hernandez, E. et al. (2017). Plastic Teabags Release Billions of Microparticles and Nanoparticles into Tea. *Environmental Science & Technology*, 53(21), 12300–12310.

22. Dong, C.D. et al. (2020). Nanoplastics crossing through skin: A new pathway? *Journal of Hazardous Materials*, 384, 121409.

23. Walczak, A.P. et al. (2013). Bioavailability and uptake of nano-plastics by gastrointestinal tract and transport to internal organs. *Environmental Science: Nano*, 2(6), 615–621.

24. Panyam, J. & Labhasetwar, V. (2003). Biodegradable nanoparticles for drug and gene delivery to cells and tissue. *Advanced Drug Delivery Reviews*, 55(3), 329–347.

25. Weir, A. et al. (2012). Use of nanotechnology in food products. *Environmental Science & Technology*, 46(4), 2242–2250.

26. Sullivan, G.L. et al. (2022). Micro- and nanoplastic exposure from disposable medical masks. *Science of the Total Environment*, 816, 151547.

27. Hahladakis, J.N. et al. (2018). An overview of the challenges in cleaning and decontaminating plastic materials in hospitals. *Waste Management*, 77, 147–160.

28. Abbasi, S. et al. (2018). Microplastics in different tissues of fish and human exposure. *Environmental Pollution*, 237, 650–659.

29. Schwabl, P. et al. (2019). Detection of various microplastics in human stool: a prospective case series. *Annals of Internal Medicine*, 171(7), 453–457.

30. Stock, V. et al. (2021). Uptake and cellular effects of microplastics in vitro: a systematic review of the current literature. *Environment International*, 146, 106274.

31. Deng, Y. et al. (2017). Nanoplastics can cross the intestinal barrier and reach surrounding tissues. *Nature Nanotechnology*, 12(12), 1121–1128.

32. Walczak, A.P. et al. (2015). Paracellular transport of polystyrene nanoparticles in vitro: effect of surface charge, size, and shape. *Toxicology In Vitro*, 29(1), 231–238.

33. Forte, M. et al. (2016). Endocytosis and intracellular fate of nanoplastics in human intestinal Caco-2 cells. *Nanotoxicology*, 10(6), 863–871.

34. Wright, S.L. & Kelly, F.J. (2017). Plastic and human health: a micro issue? *Environmental Science & Technology*, 51(12), 6634–6647.

35. Prata, J.C. (2018). Airborne microplastics: Consequences to human health? *Environmental Pollution*, 234, 115–126.

36. Vianello, A. et al. (2019). Simulating human exposure to indoor airborne microplastics using a breathing thermal manikin. *Scientific Reports*, 9, 8670.

37. Lim, X. (2021). Microplastics are everywhere — but are they harmful? *Nature*, 593(7857), 22–25.

38. Danopoulos, E. et al. (2020). Dermal exposure to microplastic particles: A potential pathway for systemic uptake. *Science of the Total Environment*, 703, 134725.

39. Mahler, G.J. et al. (2012). Dermal penetration of nanoparticles and effects on skin health. *Advanced Drug Delivery Reviews*, 64(15), 1731–1740.

40. Sharma, S. et al. (2020). Microplastics as pollutants in aquatic systems and human health: A review. *Environmental Science and Pollution Research*, 27, 15557–15564.

41. Leslie, H.A. et al. (2022). Discovery and quantification of plastic particle pollution in human placentas. *Environment International*, 146, 106228.

42. Ragusa, A. et al. (2021). Plasticenta: First evidence of microplastics in human placenta. *Environment International*, 146, 106274.

43. An, R. et al. (2021). Potential neurotoxic effects of nanoplastics and their neurotoxicity pathways. *Chemosphere*, 271, 129421.

44. Hirt, N. & Body-Malapel, M. (2020). Immunotoxicity and intestinal effects of nano- and microplastics: a review of the literature. *Particle and Fibre Toxicology*, 17, 57.

45. Yong, C.Q.Y. et al. (2020). Toxicity of microplastics and nanoplastics in mammalian systems. *International Journal of Environmental Research and Public Health*, 17(5), 1509.

46. Pitt, J.A. et al. (2018). Uptake, tissue distribution, and toxicological effects of polystyrene microplastics in developing zebrafish. *Aquatic Toxicology*, 194, 185–194.

47. Roh, J.Y. et al. (2009). Physicochemical properties of nanomaterials and their cellular effects. *Environmental and Molecular Mutagenesis*, 50(1), 23–30.

48. Bhattacharya, P. et al. (2010). Physical adsorption of charged plastic nanoparticles affects algal photosynthesis. *The Journal of Physical Chemistry C*, 114(39), 16556–16561.

49. Yang, Y. et al. (2019). Microplastic pollution and oxidative stress: a review of recent findings. *Science of the Total Environment*, 683, 610–617.

50. Hwang, J. et al. (2019). Potential toxicity of polystyrene microplastic particles. *Scientific Reports*, 9, 7391.

51. Galloway, T.S. et al. (2017). Interactions of microplastic debris throughout the marine ecosystem. *Nature Ecology & Evolution*, 1(5), 0116.

52. Chen, Q. et al. (2019). Eco-coronas determine the biological fates of microplastics. *Nature Nanotechnology*, 14(4), 300–306.

53. Kirstein, I.V. et al. (2016). Dangerous hitchhikers? Evidence for potentially pathogenic Vibrio spp. on microplastic particles. *Marine Environmental Research*, 120, 1–8.

54. Talsness, C.E. et al. (2009). Components of plastic: experimental studies in animals and relevance for human health. *Philosophical Transactions of the Royal Society B*, 364(1526), 2079–2096.

55. Gore, A.C. et al. (2015). EDC-2: The Endocrine Society's Second Scientific Statement on Endocrine-Disrupting Chemicals. *Endocrine Reviews*, 36(6), E1–E150.

56. Witorsch, R.J. & Thomas, J.A. (2010). Personal care products and endocrine disruption: A critical review. *Critical Reviews in Toxicology*, 40(S3), 1–30.

57. Koelmans, A.A. et al. (2015). Nanoplastics in the aquatic environment. Critical review. *Marine Anthropogenic Litter*, 325–340.

58. Hadiyanto, H. et al. (2022). Influence of shape and surface charge of microplastics on human health risk: A review. *Environmental Research*, 210, 112915.

59. Lithner, D. et al. (2011). Environmental and health hazard ranking and assessment of plastic polymers based on chemical composition. *Science of the Total Environment*, 409(18), 3309–3324.

60. Smith, M. et al. (2018). Microplastics in seafood and the implications for human health. *Current Environmental Health Reports*, 5(3), 375–386.

61. Campanale, C. et al. (2020). A detailed review study on potential effects of microplastics and additives of concern on human health. *International Journal of Environmental Research and Public Health*, 17(4), 1212.

62. Leslie, H.A. et al. (2022). Discovery and quantification of plastic particle pollution in human blood. *Environment International*, 163, 107199.

63. Zhang, Q. et al. (2021). Microplastics in infant feeding bottles and formula. *Nature Food*, 2(10), 735–741.

64. Gasperi, J. et al. (2018). Microplastics in air: Are we breathing it in? *Current Opinion in Environmental Science & Health*, 1, 1–5.

65. Boucher, J. & Friot, D. (2017). Primary microplastics in the oceans: a global evaluation of sources. *IUCN Reports*.

66. Prata, J.C. et al. (2020). Microplastics in occupational settings: exposure, effects, and solutions. *Environmental International*, 140, 105805.

67. Leslie, H.A. et al. (2022). Discovery and quantification of plastic particle pollution in human blood. *Environment International*, 163, 107199.

68. Ragusa, A. et al. (2021). Plasticenta: First evidence of microplastics in human placenta. *Environment International*, 146, 106274.

69. Ibrahim, Y.S. et al. (2021). Detection of microplastics in human saphenous vein tissue. *Environmental Science and Pollution Research*, 28, 36027–36035.

70. Yong, C.Q.Y. et al. (2020). Toxicity of microplastics and nanoplastics in mammalian systems: A review. *International Journal of Environmental Research and Public Health*, 17(5), 1509.

71. Schwabl, P. et al. (2019). Detection of various microplastics in human stool: a prospective case series. *Annals of Internal Medicine*, 171(7), 453–457.

72. Smith, M. et al. (2018). Microplastics in seafood and the implications for human health. *Current Environmental Health Reports*, 5(3), 375–386.

73. Koelmans, A.A. et al. (2016). Microplastic as a vector for chemicals in the aquatic environment: Critical review and model-supported reinterpretation of empirical studies. *Environmental Science & Technology*, 50(7), 3315–3326.

74. Sharma, S. & Chatterjee, S. (2017). Microplastic pollution, a threat to marine ecosystem and human health: a short review. *Environmental Science and Pollution Research*, 24(27), 21530–21547.

75. Talsness, C.E. et al. (2009). Components of plastic: experimental studies in animals and relevance for human health. *Philosophical Transactions of the Royal Society B*, 364(1526), 2079–2096.

76. Gallo, F. et al. (2018). Marine litter plastics and microplastics and their toxic chemicals components: the need for urgent preventive measures. *Environmental Sciences Europe*, 30, 13.

77. Kirstein, I.V. et al. (2016). Dangerous hitchhikers? Evidence for potentially pathogenic Vibrio spp. on microplastic particles. *Marine Environmental Research*, 120, 1–8.

78. Ragusa, A. et al. (2021). Plasticenta: First evidence of microplastics in human placenta. *Environment International*, 146, 106274.

79. Hou, L. et al. (2021). Effects of microplastics exposure on reproductive and endocrine systems in animal models. *Science of The Total Environment*, 755, 142470.

80. Hwang, J. et al. (2019). Potential toxicity of polystyrene microplastic particles. *Scientific Reports*, 9(1), 7391.

81. Leslie, H.A. et al. (2022). Discovery and quantification of plastic particle pollution in human blood. *Environment International*, 163, 107199.

82. SAPEA. (2019). A scientific perspective on microplastics in nature and society. *Science Advice for Policy by European Academies*.

83. Lithner, D. et al. (2011). Environmental and health hazard ranking and assessment of plastic polymers. *Science of the Total Environment*, 409(18), 3309–3324.

84. Barboza, L.G.A. et al. (2020). Microplastics in wild fish from North East Atlantic Ocean and its potential for causing neurotoxicity. *Environmental Pollution*, 263, 114631.

85. UNEP. (2021). Neglected: Environmental justice and plastic pollution. *United Nations Environment Programme Report*.

86. Wright, S.L. & Kelly, F.J. (2017). Plastic and human health: A micro issue? *Environmental Science & Technology*, 51(12), 6634–6647.

Chapter 6 References:

1. Hwang, J. et al. (2019). Potential toxicity of polystyrene microplastic particles. *Scientific Reports*, 9(1), 7391.

2. Prata, J.C. (2018). Microplastics in wastewater: State of the knowledge on sources, fate and solutions. *Marine Pollution Bulletin*, 129(1), 262–265.

3. Wright, S.L. et al. (2017). Plastic and human health: A micro issue? *Environmental Science & Technology*, 51(12), 6634–6647.

4. Dong, C.D. et al. (2020). Polystyrene microplastic exposure induces metabolic disturbance in mice. *Science of the Total Environment*, 726, 138689.

5. Halden, R.U. (2010). Plastics and health risks. *Reviews on Environmental Health*, 25(1), 1–24.

6. Rochester, J.R. (2013). Bisphenol A and human health: A review of the literature. *Reproductive Toxicology*, 42, 132–155.

7. Pitt, J.A. et al. (2018). Uptake, tissue distribution, and toxicity of polystyrene nanoparticles in developing zebrafish. *Aquatic Toxicology*, 194, 185–194.

8. Chen, Q. et al. (2017). Polystyrene microplastics cause liver injury by affecting the endoplasmic reticulum stress pathway. *Environmental Toxicology and Pharmacology*, 46, 62–70.

9. Vianello, A. et al. (2019). Simulating human exposure to indoor airborne microplastics using a breathing thermal manikin. *Scientific Reports*, 9, 8670.

10. Prata, J.C. et al. (2020). Solutions and integrated strategies for the control and mitigation of plastic and microplastic pollution. *International Journal of Environmental Research and Public Health*, 17(23), 9130.

11. Gasperi, J. et al. (2018). Microplastics in air: Are we breathing it in? *Current Opinion in Environmental Science & Health*, 1, 1–5.

12. Deng, Y. et al. (2017). Tissue accumulation of microplastics in mice and biomarker responses suggest widespread health risks. *Nature Communications*, 8, 15611.

13. Jin, Y. et al. (2018). Polystyrene microplastics induce microbiota dysbiosis and inflammation in mice. *Science of the Total Environment*, 630, 118–127.

14. Li, Y. et al. (2021). Kidney injury induced by microplastics exposure in mice. *Chemosphere*, 275, 130015.

15. Du, F. et al. (2020). Toxic effects of micro- and nanoplastics on mammalian kidney function. *Environmental Pollution*, 262, 114350.

16. Lu, L. et al. (2018). Uptake and accumulation of polystyrene microplastics in zebrafish and their effect on the gut microbiota. *Environmental Pollution*, 238, 1–9.

17. Wang, Y.L. et al. (2019). Effects of polystyrene microplastics on gut barrier, microbiota and metabolism in mice. *Science of the Total Environment*, 708, 135232.

18. Mattsson, K. et al. (2017). Brain damage and behavioral disorders in fish induced by plastic nanoparticles delivered through the food chain. *Scientific Reports*, 7, 11452.

19. Hou, L. et al. (2021). Microplastics exposure and neurotoxicity: A review of current evidence. *Environmental Pollution*, 274, 116569.

20. Meeker, J.D. et al. (2009). Environmental exposure to metals and male reproductive hormones: The NHANES study. *Environmental Health Perspectives*, 118(2), 297–302.

21. Buck Louis, G.M. et al. (2013). Persistent environmental pollutants and couple fecundity: The LIFE study. *Environmental Health Perspectives*, 121(2), 231–236.

22. Hunt, P.A. et al. (2016). Plastic-derived chemicals and reproductive health. *Nature Reviews Endocrinology*, 12(10), 537–546.

23. Ragusa, A. et al. (2021). Plastic particles in human placentas. *Environment International*, 146, 106274.

24. Hou, L. et al. (2022). Prenatal microplastic exposure causes neurodevelopmental toxicity in mice offspring. *Journal of Hazardous Materials*, 422, 126885.

25. Boudou, A. et al. (2020). Epigenetic alterations following microplastic exposure: Emerging concern for transgenerational effects. *Toxicology Letters*, 331, 13–20.

26. Schirinzi, G.F. et al. (2017). Cytotoxic effects of micro- and nanoplastics on human cells. *Environmental Research*, 159, 579–587.

27. D'Angelo, S. et al. (2021). Nanoplastics exposure and immune system disruption: Mechanisms and models. *Frontiers in Toxicology*, 3, 742356.

28. Forte, M. et al. (2021). Autoimmunity and environmental pollutants: Nanoplastics as emerging triggers. *International Journal of Environmental Research and Public Health*, 18(15), 7811.

29. Wu, J.H. et al. (2019). Polystyrene microplastic exposure induces inflammation and oxidative stress. *Chemosphere*, 222, 714–723.

30. Hirt, N. & Body-Malapel, M. (2020). Immunotoxicity of plastic particles: A growing concern. *Toxics*, 8(2), 23.

31. Prata, J.C. et al. (2021). Environmental exposure to plastic and allergic skin conditions: Is there a link? *Allergology International*, 70(3), 292–299.

32. Rubin, B.S. (2011). Bisphenol A: An endocrine disruptor with widespread exposure and multiple effects. *Journal of Steroid Biochemistry and Molecular Biology*, 127(1-2), 27–34.

33. Stahlhut, R.W. et al. (2007). Association of phthalate exposure with low testosterone in men. *Environmental Health Perspectives*, 115(6), 876–882.

34. Janesick, A.S. & Blumberg, B. (2016). Obesogens: An emerging threat to public health. *American Journal of Obstetrics and Gynecology*, 214(5), 559–565.

35. Lim, S. et al. (2009). Chronic exposure to BPA induces glucose intolerance and insulin resistance. *Journal of Endocrinology*, 202(3), 385–392.

36. Zhang, H. et al. (2020). Microplastic exposure alters gut microbiota and promotes obesity. *Science of the Total Environment*, 715, 136990.

37. Kim, K.H. et al. (2021). Endocrine-disrupting chemicals and blood pressure regulation. *Environmental Research*, 198, 111222.

38. D'Amato, G. et al. (2020). Inflammatory pathways in cardiovascular disease and plastic exposure. *International Journal of Environmental Research and Public Health*, 17(14), 5076.

39. Leslie, H.A. et al. (2022). Discovery and implications of microplastics in human blood. *Environmental International*, 163, 107199.

40. Ibid.

41. Wright, S.L. & Kelly, F.J. (2017). Plastic and human health: A micro issue? *Environmental Science & Technology*, 51(12), 6634–6647.

42. Deng, Y. et al. (2018). Nanoplastics and human health: Interaction with the blood-brain barrier. *Science of the Total Environment*, 621, 1373–1379.

43. Prüst, M. et al. (2020). Plastic nanoparticles induce neurotoxicity through microglial activation. *Toxicology and Applied Pharmacology*, 387, 114846.

44. Chen, Q. et al. (2021). Behavioral and memory impairment in mice following microplastic ingestion. *Chemosphere*, 263, 128164.

45. Zhang, Y. et al. (2020). Dopaminergic and serotonergic modulation by nanoplastics. *Ecotoxicology and Environmental Safety*, 206, 111108.

46. Harley, K.G. et al. (2013). Prenatal BPA exposure and child neurodevelopment. *Environmental Health Perspectives*, 121(4), 514–520.

47. Braun, J.M. et al. (2011). BPA and behavioral outcomes in children. *Pediatrics*, 128(5), 873–882.

48. Vuong, A.M. et al. (2020). Prenatal BPA and autism spectrum behaviors. *Environmental Research*, 181, 108727.

49. Barone, I. et al. (2021). Endocrine disruption and circadian rhythm disturbances. *Endocrine Reviews*, 42(4), 590–610.

50. IARC Monographs. (2021). List of Classifications by Cancer Site. *International Agency for Research on Cancer.*

51. Sharma, S. & Chatterjee, S. (2017). Genotoxic effects of microplastics. *Environmental Toxicology and Pharmacology*, 45, 239–247.

52. Lu, Y. et al. (2016). Liver injury in mice from polystyrene microplastics. *Scientific Reports*, 6, 33935.

53. Koelmans, A.A. et al. (2016). Microplastics as vectors for chemicals. *Environmental Science & Technology*, 50(7), 3315–3326.

54. Campanale, C. et al. (2022). Nanoplastics and the tumor microenvironment. *Nanotoxicology*, 16(2), 146–158.

55. Diamanti-Kandarakis, E. et al. (2009). Endocrine-disrupting chemicals: An Endocrine Society scientific statement. *Endocrine Reviews*, 30(4), 293–342.

56. Seachrist, D.D. et al. (2016). BPA and breast cancer risk. *Reproductive Toxicology*, 59, 167–186.

57. Zhang, W. et al. (2022). Exposure risks for children in plastic-rich indoor environments. *Building and Environment*, 208, 108583.

58. Ragusa, A. et al. (2021). Plastic particles in human placentas: First detection and implications. *Environment International*, 146, 106274.

59. Lavelle, M. et al. (2020). Age-related differences in detox pathways and plastic exposure. *Geriatric Medicine Journal*, 25(3), 211–218.

60. Prata, J.C. et al. (2020). Microplastic inhalation and occupational risk. *Environmental Pollution*, 265, 114940.

61. Taylor, M.P. et al. (2019). Socioeconomic disparities and plastic pollution exposure. *Nature Sustainability*, 2(7), 561–568.

62. Morello-Frosch, R. et al. (2011). Environmental justice and cumulative risk assessment. *Annual Review of Public Health*, 32, 103–124.

63. Holmes, L.A. et al. (2012). Adsorption of trace metals to plastic resin pellets in the marine environment. *Marine Pollution Bulletin*, 60(11), 2003–2008.

64. Wang, J. et al. (2016). Transport of heavy metals by microplastics in freshwater systems. *Environmental Pollution*, 219, 1–7.

65. Bakir, A. et al. (2014). Enhanced desorption of persistent organic pollutants from microplastics under simulated physiological conditions. *Environmental Pollution*, 185, 16–23.

66. Chua, E.M. et al. (2014). Combined effects of microplastics and persistent organic pollutants in aquatic organisms. *Marine Environmental Research*, 96, 61–68.

67. Zhang, Q. et al. (2021). Microplastics as carriers of antibiotic resistance genes in wastewater. *Science of the Total Environment*, 760, 144163.

68. Yang, Y. et al. (2019). Potential risks posed by microplastics in wastewater treatment plants. *Journal of Hazardous Materials*, 365, 715–723.

69. Prata, J.C. et al. (2019). Methods for detecting microplastics in human matrices: A critical review. *TrAC Trends in Analytical Chemistry*, 110, 150–159.

70. Yong, C.Q.Y. et al. (2020). Toxicity of microplastics and nanoplastics in mammalian systems. *International Journal of Environmental Research and Public Health*, 17(5), 1509.

71. Vethaak, A.D. & Leslie, H.A. (2016). Plastic debris as a human health hazard. *Integrated Environmental Assessment and Management*, 12(3), 346–349.

72. Hahladakis, J.N. & Iacovidou, E. (2018). An overview of legislation and policies addressing plastics and microplastics pollution. *Science of the Total Environment*, 643, 785–795.

73. Landrigan, P.J. & Fuller, R. (2015). Environmental pollution: An under-recognized cause of non-communicable diseases. *The Lancet*, 386(10003), 1429–1430.

74. Koelmans, A.A. et al. (2021). Assessing the risks of microplastics to human health: Current knowledge and future priorities. *Environmental Toxicology and Chemistry*, 40(1), 9–15.

75. Singh, B. et al. (2021). Microbial and enzymatic degradation of synthetic plastics. *Nature Reviews Microbiology*, 19(12), 803–816.

76. Galloway, T.S. & Lewis, C.N. (2016). Marine microplastics spell big problems for future generations. *Proceedings of the National Academy of Sciences*, 113(9), 2331–2333.

77. Bullard, R.D. (1993). Race and environmental justice in the United States. *Yale Journal of International Law*, 18(1), 319–335.

Chapter 7 References:

1. de Souza Machado, A. A., et al. (2018). Impacts of microplastics on the soil biophysical environment. *Environmental Science & Technology*, 52(17), 9656–9665.

2. Rillig, M. C., Ziersch, L., & Hempel, S. (2017). Microplastic transport in soil by earthworms. *Scientific Reports*, 7, 1362.

3. Ng, E. L., et al. (2018). An overview of microplastic and nanoplastic pollution in agroecosystems. *Science of The Total Environment*, 627, 1377–1388.

4. Kögel, T., et al. (2020). Micro- and nanoplastics in marine species: from the coastline to the fish gut. *Environmental Pollution*, 265, 114481.

5. Besseling, E., et al. (2014). Nanoplastic affects growth of *S. obliquus* and reproduction of *D. magna*. *Environmental Science & Technology*, 48(20), 12336–12343.

6. Cole, M., et al. (2011). Microplastics as contaminants in the marine environment: a review. *Marine Pollution Bulletin*, 62(12), 2588–2597.

7. Andrady, A. L. (2011). Microplastics in the marine environment. *Marine Pollution Bulletin*, 62(8), 1596–1605.

8. Van Cauwenberghe, L., et al. (2015). Microplastics in bivalves cultured for human consumption. *Environmental Pollution*, 199, 10–15.

9. Wright, S. L., Thompson, R. C., & Galloway, T. S. (2013). The physical impacts of microplastics on marine organisms. *Environmental Pollution*, 178, 483–492.

10. Lamb, J. B., et al. (2018). Plastic waste associated with disease on coral reefs. *Science*, 359(6374), 460–462.

11. Hall, N. M., et al. (2015). Microplastic ingestion by scleractinian corals. *Marine Biology*, 162, 725–732.

12. Romeo, T., et al. (2015). First evidence of presence of plastic debris in stomach of large pelagic fish. *Marine Pollution Bulletin*, 95(1), 358–361.

13. Fossi, M. C., et al. (2014). Are baleen whales exposed to the threat of microplastics? *Marine Pollution Bulletin*, 86(1–2), 258–264.

14. Provencher, J. F., et al. (2015). Quantifying ingested debris in marine megafauna. *Marine Pollution Bulletin*, 93(1–2), 159–164.

15. Wilcox, C., et al. (2015). Threat of plastic pollution to seabirds is global, pervasive, and increasing. *PNAS*, 112(38), 11899–11904.

16. Acampora, H., et al. (2014). Plastic ingestion by seabird chicks: patterns and consequences. *Marine Pollution Bulletin*, 78(1), 18–24.

17. Prata, J. C., et al. (2020). Environmental exposure to microplastics: an overview on possible human health effects. *Science of The Total Environment*, 702, 134455.

18. Zhang, Y., et al. (2022). Uptake and effects of microplastics in terrestrial insects: evidence and knowledge gaps. *Science of The Total Environment*, 806, 150621.

19. Boyle, M., et al. (2021). The potential impact of microplastics on amphibians and reptiles. *Environmental Pollution*, 270, 116261.

20. Huerta B., et al. (2019). Microplastic ingestion by amphibian larvae. *Ecotoxicology*, 28(4), 529–537.

21. Oehlmann, J., et al. (2009). A critical analysis of endocrine disruption in wildlife. *Environmental Research*, 109(8), 926–940.

22. Pompini, M., et al. (2022). Developmental deformities in amphibians linked to microplastic exposure. *Scientific Reports*, 12, 6124.

23. Klein, M., et al. (2020). Environmental stressors in amphibian breeding ponds. *Aquatic Toxicology*, 227, 105618.

24. Wake, D. B., & Vredenburg, V. T. (2008). Are we in the midst of the sixth mass extinction? *PNAS*, 105(Suppl 1), 11466–11473.

25. Schuyler, Q., et al. (2014). Risk analysis of plastic ingestion by sea turtles. *Conservation Biology*, 28(1), 129–139.

26. Gili, C., et al. (2021). Microplastic exposure in terrestrial reptiles. *Environmental Pollution*, 279, 116865.

27. Alava, J. J., et al. (2022). Plastic pollution impacts on reptile reproduction. *Global Change Biology*, 28(10), 3063–3078.

28. Brühl, C. A., et al. (2013). Terrestrial pesticide exposure of amphibians. *Environmental Toxicology and Chemistry*, 32(3), 647–655.

29. Urban, R., et al. (2021). Plastic use in solitary bee nests. *Environmental Science & Technology*, 55(14), 9796–9804.

30. Sánchez-Bayo, F., & Wyckhuys, K. A. (2019). Worldwide decline of the entomofauna. *Biological Conservation*, 232, 8–27.

31. de Souza Machado, A. A., et al. (2019). Microplastics as an emerging threat to terrestrial ecosystems. *Science of The Total Environment*, 649, 1004–1015.

32. Potts, S. G., et al. (2010). Global pollinator declines. *Trends in Ecology & Evolution*, 25(6), 345–353.

33. Lwanga, E. H., et al. (2016). Microplastics reduce litter decomposition by soil fauna. *Scientific Reports*, 6, 38069.

34. Leifheit, E. F., et al. (2021). Impact of microplastics on mycorrhizal fungi. *Frontiers in Plant Science*, 12, 661728.

35. Woodall, L. C., et al. (2014). The deep sea is a major sink for microplastic debris. *Royal Society Open Science*, 1(4), 140317.

36. Rillig, M. C., & Lehmann, A. (2020). Microplastic effects on soil functions. *Nature Reviews Earth & Environment*, 1(6), 294–303.

37. Farrell, P., & Nelson, K. (2013). Trophic level transfer of microplastic. *Environmental Pollution*, 177, 1–3.

38. Estes, J. A., et al. (2011). Trophic downgrading of planet Earth. *Science*, 333(6040), 301–306.

39. Courtene-Jones, W., et al. (2017). Microplastics in migratory species. *Marine Pollution Bulletin*, 117(1–2), 283–291.

40. Rohr, J. R., et al. (2008). Multiple stressors and population regulation. *Ecology Letters*, 11(11), 1304–1315.

41. Goldstein, M. C., et al. (2012). Microplastic ingestion by zooplankton. *Marine Ecology Progress Series*, 426, 129–142.

42. Davison, P., & Asch, R. G. (2011). Microplastic ingestion by mesopelagic fishes. *Marine Ecology Progress Series*, 432, 173–180.

43. Tanaka, K., et al. (2013). Accumulation of plastic-derived chemicals in seabirds. *Marine Pollution Bulletin*, 69(1–2), 219–222.

44. Choy, C. A., et al. (2019). The vertical distribution and biological transport of marine microplastics. *Nature Scientific Reports*, 9, 7843.

45. Horton, A. A., Walton, A., Spurgeon, D. J., Lahive, E., & Svendsen, C. (2017). Microplastics in freshwater and terrestrial environments: Evaluating the current understanding to identify the knowledge gaps. *Science of The Total Environment*, 586, 127–141.

46. Windsor, F. M., Tilley, R. M., Tyler, C. R., & Ormerod, S. J. (2019). Microplastic ingestion by riverine macroinvertebrates. *Science of The Total Environment*, 646, 68–74.

47. Nelms, S. E., Galloway, T. S., Godley, B. J., Jarvis, D. S., & Lindeque, P. K. (2018). Investigating microplastic trophic transfer in marine top predators. *Environmental Pollution*, 238, 999–1007.

48. Huang, Y., Liu, Q., Jia, W., Yan, C., & Wang, J. (2020). Agricultural plastic mulch film residues in farmland soils and their ecological risks. *Science of The Total Environment*, 651, 2284–2293.

49. Liu, E. K., He, W. Q., & Yan, C. R. (2014). 'White revolution' to 'white pollution'—agricultural plastic film mulch in China. *Environmental Research Letters*, 9(9), 091001.

50. Rillig, M. C., & Lehmann, A. (2020). Microplastic in terrestrial ecosystems. *Science*, 368(6498), 1430–1431.

51. Jambeck, J. R., Geyer, R., Wilcox, C., et al. (2015). Plastic waste inputs from land into the ocean. *Science*, 347(6223), 768–771.

52. The Ocean Cleanup. (2020). Interceptor solutions for river plastic waste. Retrieved from https://theoceancleanup.com/interceptor/

53. Rochman, C. M., Brookson, C., Bikker, J., et al. (2019). Rethinking microplastics as a diverse contaminant suite. *Environmental Toxicology and Chemistry*, 38(4), 703–711.

54. Yoshida, S., Hiraga, K., Takehana, T., et al. (2016). A bacterium that degrades and assimilates poly(ethylene terephthalate). *Science*, 351(6278), 1196–1199.

55. Shah, A. A., Hasan, F., Hameed, A., & Ahmed, S. (2008). Biological degradation of plastics: A comprehensive review. *Biotechnology Advances*, 26(3), 246–265.

56. Danso, D., Chow, J., & Streit, W. R. (2019). Plastics: Environmental and biotechnological perspectives on microbial degradation. *Applied and Environmental Microbiology*, 85(19), e01095–19.

57. Sala, E., Giakoumi, S., & Friedlander, A. M. (2018). No-take marine reserves are the most effective protected areas in the ocean. *ICES Journal of Marine Science*, 75(3), 1166–1168.

58. Xanthos, D., & Walker, T. R. (2017). International policies to reduce plastic marine pollution from single-use plastics (plastic bags and microbeads): A review. *Marine Pollution Bulletin*, 118(1–2), 17–26.

59. Basel Convention on the Control of Transboundary Movements of Hazardous Wastes and Their Disposal. (2019). United Nations Environment Programme.

Chapter 8 References:

1. Lebreton, L. C., Slat, B., Ferrari, F., et al. (2018). Evidence that the Great Pacific Garbage Patch is rapidly accumulating plastic. *Scientific Reports*, 8(1), 4666.

2. Andrady, A. L. (2011). Microplastics in the marine environment. *Marine Pollution Bulletin*, 62(8), 1596–1605.

3. Choy, C. A., Robison, B. H., Gagne, T. O., et al. (2019). The vertical distribution and biological transport of marine microplastics across the epipelagic and mesopelagic water column. *Scientific Reports*, 9, 7843.

4. Eriksen, M., Lebreton, L. C., Carson, H. S., et al. (2014). Plastic pollution in the world's oceans: More than 5 trillion plastic pieces weighing over 250,000 tons afloat at sea. *PLOS ONE*, 9(12), e111913.

5. Schmidt, C., Krauth, T., & Wagner, S. (2017). Export of plastic debris by rivers into the sea. *Environmental Science & Technology*, 51(21), 12246–12253.

6. Lebreton, L. M., Greer, S. D., & Borrero, J. C. (2012). Numerical modelling of floating debris in the world's oceans. *Marine Pollution Bulletin*, 64(3), 653–661.

7. Helmholtz Centre for Environmental Research. (2017). Study: Rivers carry plastic into the sea. *ScienceDaily*.

8. Jambeck, J. R., Geyer, R., Wilcox, C., et al. (2015). Plastic waste inputs from land into the ocean. *Science*, 347(6223), 768–771.

9. Mintenig, S. M., Int-Veen, I., Löder, M. G. J., Primpke, S., & Gerdts, G. (2017). Identification of microplastic in effluents of waste water treatment plants using focal plane array-based micro-Fourier-transform infrared imaging. *Water Research*, 108, 365–372.

10. Eriksen, M., Mason, S., Wilson, S., et al. (2013). Microplastic pollution in the surface waters of the Laurentian Great Lakes. *Marine Pollution Bulletin*, 77(1–2), 177–182.

11. Faure, F., Corbaz, M., Baecher, H., & de Alencastro, L. F. (2012). Pollution due to plastics and microplastics in Lake Geneva and in the Mediterranean Sea. *Archives des Sciences*, 65, 157–164.

12. Lagarde, F., Olivier, O., Zanella, M., et al. (2016). Microplastic interactions with freshwater microalgae: Hetero-aggregation and changes in plastic density appear strongly dependent on polymer type. *Environmental Pollution*, 215, 331–339.

13. Zhang, G. S., & Liu, Y. F. (2018). The distribution of microplastics in soil aggregate fractions in southwestern China. *Science of The Total Environment*, 642, 12–20.

14. Yu, H., Huang, Y., Zhang, W., et al. (2021). Impact of plastic film residues on soil physical and chemical properties in farmlands of China: A meta-analysis. *Science of The Total Environment*, 780, 146583.

15. Wang, J., Liu, X., Li, Y., et al. (2020). Microplastics as contaminants in the soil environment: A mini-review. *Science of The Total Environment*, 691, 848–857.

16. Carr, S. A., Liu, J., & Tesoro, A. G. (2016). Transport and fate of microplastic particles in wastewater treatment plants. *Water Research*, 91, 174–182.

17. Murphy, F., Ewins, C., Carbonnier, F., & Quinn, B. (2016). Wastewater treatment works (WwTW) as a source of microplastics in the aquatic environment. *Environmental Science & Technology*, 50(11), 5800–5808.

18. Horton, A. A., Walton, A., Spurgeon, D. J., Lahive, E., & Svendsen, C. (2017). Microplastics in freshwater and terrestrial environments: Evaluating the current understanding to identify the knowledge gaps and future research priorities. *Science of The Total Environment*, 586, 127–141.

19. Geyer, R., Jambeck, J. R., & Law, K. L. (2017). Production, use, and fate of all plastics ever made. *Science Advances*, 3(7), e1700782.

20. He, D., Luo, Y., Lu, S., Liu, M., Song, Y., & Lei, L. (2018). Microplastics in soils: Analytical methods, pollution characteristics and ecological risks. *TrAC Trends in Analytical Chemistry*, 109, 163–172.

21. Rochman, C. M., Browne, M. A., Halpern, B. S., et al. (2013). Policy: Classify plastic waste as hazardous. *Nature*, 494(7436), 169–171.

22. Dris, R., Gasperi, J., Saad, M., Mirande, C., & Tassin, B. (2016). Synthetic fibers in atmospheric fallout: A source of microplastics in the environment? *Marine Pollution Bulletin*, 104(1–2), 290–293.

23. Zhang, Y., Kang, S., Allen, S., et al. (2020). Atmospheric microplastics: A review on current status and perspectives. *Earth-Science Reviews*, 203, 103118.

24. Prata, J. C. (2018). Airborne microplastics: Consequences to human health? *Environmental Pollution*, 234, 115–126.

25. Allen, S., Allen, D., Phoenix, V. R., et al. (2019). Atmospheric transport and deposition of microplastics in a remote mountain catchment. *Nature Geoscience*, 12, 339–344.

26. Brahney, J., Mahowald, N., Prank, M., et al. (2021). Constraining the atmospheric limb of the plastic cycle. *Proceedings of the National Academy of Sciences*, 118(16), e2020719118.

27. Bergmann, M., Mützel, S., Primpke, S., et al. (2019). White and wonderful? Microplastics prevail in snow from the Alps to the Arctic. *Science Advances*, 5(8), eaax1157.

28. Li, J., Qu, X., Su, L., et al. (2016). Microplastics in mussels along the coastal waters of China. *Environmental Pollution*, 214, 177–184.

29. Van Cauwenberghe, L., Claessens, M., Vandegehuchte, M. B., & Janssen, C. R. (2015). Microplastics are taken up by mussels (Mytilus edulis) and lugworms (Arenicola marina) living in natural habitats. *Environmental Pollution*, 199, 10–17.

30. Santana, M. F. M., Ascer, L. G., Custódio, M. R., et al. (2016). Microplastic contamination in natural mussel beds from a Brazilian urbanized coastal region: Rapid evaluation through bioassessment. *Marine Pollution Bulletin*, 106(1–2), 183–189.

31. Liu, F., Olesen, K. B., Borregaard, A. R., & Vollertsen, J. (2019). Microplastics in urban and highway stormwater retention ponds. *Science of The Total Environment*, 671, 992–1000.

32. Karami, A., Golieskardi, A., Keong Choo, C., et al. (2017). Microplastic and mesoplastic contamination in canned fish. *Environmental Science & Pollution Research*, 24, 19143–19150.

33. Yang, D., Shi, H., Li, L., et al. (2015). Microplastic pollution in table salts from China. *Environmental Science & Technology*, 49(22), 13622–13627.

34. Kosuth, M., Mason, S. A., & Wattenberg, E. V. (2018). Anthropogenic contamination of tap water, beer, and sea salt. *PLOS ONE*, 13(4), e0194970.

35. Schymanski, D., Goldbeck, C., Humpf, H.-U., & Fürst, P. (2018). Analysis of microplastics in water by micro-Raman spectroscopy: Release of plastic particles from different packaging into mineral water. *Water Research*, 129, 154–162.

36. Mason, S. A., Welch, V. G., & Neratko, J. (2018). Synthetic polymer contamination in bottled water. *Frontiers in Chemistry*, 6, 407.

37. Akhbarizadeh, R., Moore, F., Keshavarzi, B., & Busquets, R. (2021). Microplastics and potentially toxic elements in coastal sediments of Iran: Impacts of urbanization and industrialization. *Environmental Pollution*, 268, 115738.

38. Kataria, P., Sharma, A., & Arora, M. (2021). Presence of microplastics in cow milk: A first report. *Journal of Hazardous Materials Letters*, 2, 100020.

39. Cox, K. D., Covernton, G. A., Davies, H. L., et al. (2019). Human consumption of microplastics. *Environmental Science & Technology*, 53(12), 7068–7074.

40. Campanale, C., Massarelli, C., Savino, I., et al. (2020). A detailed review study on potential effects of microplastics and additives of concern on human health. *International Journal of Environmental Research and Public Health*, 17(4), 1212.

41. Rist, S., & Hartmann, N. B. (2018). Aquatic ecotoxicity of microplastics and nanoplastics: Lessons learned from engineered nanomaterials. *Handbook of Environmental Chemistry*, 58, 25–49.

42. Zhang, Y., Kang, S., Allen, S., et al. (2020). Atmospheric microplastics: A review on current status and perspectives. *Earth-Science Reviews*, 203, 103118.

43. Wright, S. L., Ulke, J., Font, A., et al. (2020). Atmospheric microplastic deposition in an urban environment and an evaluation of transport. *Environment International*, 136, 105411.

44. Dris, R., Gasperi, J., Rocher, V., et al. (2015). Microplastic contamination in an urban area: A case study in Greater Paris. *Environmental Chemistry*, 12(5), 592–599.

45. Prata, J. C., da Costa, J. P., Lopes, I., et al. (2020). Environmental exposure to microplastics: An overview on possible human health effects. *Science of The Total Environment*, 702, 134455.

46. Piehl, S., Leibner, A., Löder, M. G., et al. (2018). Identification and quantification of macro- and microplastics on an agricultural farmland. *Scientific Reports*, 8(1), 17950.

47. Zhang, D., Liu, H., & Shi, H. (2020). Urban microplastic pollution: Insights into spatial variability and sources. *Environmental Pollution*, 261, 114245.

48. de Souza Machado, A. A., Kloas, W., Zarfl, C., et al. (2018). Microplastics as an emerging threat to terrestrial ecosystems. *Global Change Biology*, 24(4), 1405–1416.

49. Woodall, L. C., Sanchez-Vidal, A., Canals, M., et al. (2014). The deep sea is a major sink for microplastic debris. *Royal Society Open Science*, 1(4), 140317.

50. Jamieson, A. J., Brooks, L. S., Reid, W. D., et al. (2019). Microplastics and synthetic particles ingested by deep-sea amphipods in six of the deepest marine ecosystems on Earth. *Royal Society Open Science*, 6(2), 180667.

51. Chiba, S., Saito, H., Fletcher, R., et al. (2018). Human footprint in the abyss: 30-year records of deep-sea plastic debris. *Marine Policy*, 96, 204–212.

52. Salazar, A., et al. (2021). Indigenous perspectives on environmental stewardship in the Amazon Basin. *Global Environmental Change*, 68, 102262.

53. Halpern, B. S., et al. (2008). A global map of human impact on marine ecosystems. *Science*, 319(5865), 948–952.

54. Sheppard, C. R. C., et al. (2012). Reefs and islands of the Chagos Archipelago, Indian Ocean: Why it is the world's largest no-take marine protected area. *Aquatic Conservation*, 22(2), 232–261.

55. Allen, S., Allen, D., Phoenix, V. R., et al. (2020). Atmospheric transport and deposition of microplastics in a remote mountain catchment. *Nature Geoscience*, 13(5), 339–344.

56. Wagner, M., & Lambert, S. (Eds.). (2018). *Freshwater microplastics: Emerging environmental contaminants?* Springer.

57. Klein, M., & Fischer, E. K. (2019). Microplastic abundance in atmospheric deposition within the Metropolitan area of Hamburg, Germany. *Science of The Total Environment*, 685, 96–103.

58. United Nations Environment Programme (UNEP). (2021). *World Environment Situation Room*. Retrieved from https://wesr.unep.org

59. Lebreton, L., & Andrady, A. (2019). Future scenarios of global plastic waste generation and disposal. *Palgrave Communications*, 5, 6.

60. Barrows, A. P., et al. (2017). Grab vs. neuston tow net: A microplastic sampling performance comparison and possible advances in the field. *Analytical Methods*, 9(9), 1446–1453.

Chapter 9 References:

1. Leslie, H. A., et al. (2022). Discovery and quantification of plastic particle pollution in human blood. *Environment International*, 163, 107199.
2. Cox, K. D., et al. (2019). Human consumption of microplastics. *Environmental Science & Technology*, 53(12), 7068–7074.
3. Mason, S. A., et al. (2018). Synthetic polymer contamination in bottled water. *Frontiers in Chemistry*, 6, 407.
4. Wright, S. L., & Kelly, F. J. (2017). Plastic and human health: A micro issue? *Environmental Science & Technology*, 51(12), 6634–6647.
5. Ng, C. A., & von Goetz, N. (2016). The role of dermal absorption in risk assessment of chemicals. *Environmental Health Perspectives*, 124(7), 1164–1170.
6. Zimmermann, L., et al. (2020). Are bioplastics and plant-based materials safer than conventional plastics? *Environment International*, 145, 106066.
7. Pivokonsky, M., et al. (2018). Occurrence of microplastics in raw and treated drinking water. *Science of The Total Environment*, 643, 1644–1551.
8. Mason, S. A., Welch, V. G., & Neratko, J. (2018). Synthetic polymer contamination in bottled water. *Frontiers in Chemistry*, 6, 407.
9. van Sebille, E., et al. (2015). A global inventory of small floating plastic debris. *Environmental Research Letters*, 10(12), 124006.
10. Groh, K. J., et al. (2019). Overview of known plastic packaging-associated chemicals and their hazards. *Science of The Total Environment*, 651, 3253–3268.
11. Rüegg, J., & Govarts, E. (2021). Endocrine-disrupting chemicals: Key characteristics and current strategies. *Reproductive Toxicology*, 98, 12–20.
12. Henry, B., et al. (2019). Microfiber pollution and the textile industry: What scientists know so far. *Environmental Science & Technology*, 53(20), 11610–11613.
13. Napper, I. E., & Thompson, R. C. (2016). Release of synthetic microplastic fibers from domestic washing machines: Effects of fabric type and washing conditions. *Marine Pollution Bulletin*, 112(1–2), 39–45.

14. De Falco, F., et al. (2020). Evaluation of microplastic release from the washing of synthetic fabrics using a pilot scale washing machine. *Environmental Science and Pollution Research*, 27, 10775–10784.

15. Prata, J. C. (2018). Airborne microplastics: Consequences to human health? *Environmental Pollution*, 234, 115–126.

16. Hahladakis, J. N., & Iacovidou, E. (2018). An overview of chemical additives present in plastics: Migration, release, fate and environmental impact during their use, disposal and recycling. *Journal of Hazardous Materials*, 344, 179–199.

17. Rodrigues, M. O., et al. (2019). Microplastics: A review of the consequences to the environment and public health. *Environmental Pollution*, 242, 386–395.

18. Campanale, C., et al. (2020). A detailed review study on potential effects of microplastics and additives of concern on human health. *International Journal of Environmental Research and Public Health*, 17(4), 1212.

19. Sathyanarayana, S., et al. (2008). Baby care products: Possible sources of infant phthalate exposure. *Pediatrics*, 121(2), e260–e268.

20. Higdon, J. V., & Frei, B. (2006). Cruciferous vegetables and human cancer risk: Epidemiologic evidence and mechanistic basis. *Pharmacological Research*, 55(3), 224–236.

21. Abenavoli, L., et al. (2010). Milk thistle in liver diseases: Past, present, future. *Phytotherapy Research*, 24(10), 1423–1432.

22. Stanhope, K. L. (2016). Sugar consumption, metabolic disease and obesity: The state of the controversy. *Critical Reviews in Clinical Laboratory Sciences*, 53(1), 52–67.

23. Popkin, B. M., D'Anci, K. E., & Rosenberg, I. H. (2010). Water, hydration, and health. *Nutrition Reviews*, 68(8), 439–458.

24. Palmer, B. F., & Clegg, D. J. (2016). Electrolyte and acid–base disturbances in patients with diabetes mellitus. *New England Journal of Medicine*, 374(19), 1839–1848.

25. Whelton, A. (1999). Nephrotoxicity of nonsteroidal anti-inflammatory drugs: Physiologic foundations and clinical implications. *American Journal of Medicine*, 106(5B), 13S–24S.

26. Gibson, G. R., & Roberfroid, M. B. (1995). Dietary modulation of the human colonic microbiota: Introducing the concept of prebiotics. *Journal of Nutrition*, 125(6), 1401–1412.

27. Slavin, J. L. (2013). Fiber and prebiotics: Mechanisms and health benefits. *Nutrients*, 5(4), 1417–1435.

28. Müller-Lissner, S. A. (1988). Effect of wheat bran on weight of stool and gastrointestinal transit time: A meta-analysis. *BMJ*, 296(6622), 615–617.

29. Crinnion, W. J. (2011). Sauna as a valuable clinical tool for cardiovascular, autoimmune, toxicant-induced and other chronic health problems. *Alternative Medicine Review*, 16(3), 215–225.

30. Vodder, B. (1986). *Dr. Vodder's Manual Lymph Drainage: A Practical Guide*. Thieme.

31. Whelton, A. J., & Speth, T. F. (2008). Epsom salt and detox baths: Traditional uses and modern insights. *Integrative Medicine*, 7(6), 48–53.

32. Schraufnagel, D. E., et al. (2019). Air pollution and noncommunicable diseases: A review. *Chest*, 155(2), 409–416.

33. Brown, R. P., & Gerbarg, P. L. (2005). Sudarshan Kriya yogic breathing in the treatment of stress, anxiety, and depression: Part II—clinical applications and guidelines. *Journal of Alternative and Complementary Medicine*, 11(4), 711–717.

34. Mukherjee, P. K., et al. (2005). Medicinal plants with potential anti-inflammatory properties: A review. *Phytotherapy Research*, 19(9), 798–805.

35. Lykkesfeldt, J., & Poulsen, H. E. (2010). Is vitamin C supplementation beneficial? Lessons learned from randomised controlled trials. *British Journal of Nutrition*, 103(9), 1251–1259.

36. Traber, M. G., & Stevens, J. F. (2011). Vitamins C and E: Beneficial effects from a mechanistic perspective. *Free Radical Biology and Medicine*, 51(5), 1000–1013.

37. Scalbert, A., & Williamson, G. (2000). Dietary intake and bioavailability of polyphenols. *Journal of Nutrition*, 130(8), 2073S–2085S.

38. Kajla, P., Sharma, A., & Sood, D. R. (2015). Flaxseed—a potential functional food source. *Journal of Food Science and Technology*, 52(4), 1857–1871.

39. Toxnet (2018). Activated charcoal: Hazardous substance fact sheet. *National Library of Medicine*.

40. Khandelwal, S., et al. (2008). Potential of spirulina and chlorella in detoxification. *Indian Journal of Medical Research*, 128(5), 613–621.

41. Higdon, J. V., & Frei, B. (2006). Cruciferous vegetables and human cancer risk: Epidemiologic evidence and mechanistic basis. *Pharmacological Research*, 55(3), 224–236.

42. Simopoulos, A. P. (1999). Essential fatty acids in health and chronic disease. *American Journal of Clinical Nutrition*, 70(3), 560S–569S.

43. Kharasch, E. D. (2011). The liver's role in estrogen metabolism. *Hepatology*, 54(2), 746–748.

44. Plastic Pollution Coalition. (2020). How to do a personal plastic audit.

45. Greenpeace. (2019). *Practical Guide to Plastic-Free Living*.

46. Dodson, R. E., et al. (2012). Endocrine disruptors and air fresheners: Phthalate exposure from fragranced products. *Environmental Health Perspectives*, 120(7), 935–943.

47. Rudel, R. A., et al. (2011). Food packaging and bisphenol A and phthalate exposure. *Journal of Exposure Science & Environmental Epidemiology*, 21(5), 394–402.

48. Yang, C. Z., et al. (2011). Most plastic products release estrogenic chemicals: A potential health problem that can be solved. *Environmental Health Perspectives*, 119(7), 989–996.

49. Geueke, B. (2014). Dossier: Can coatings. *Food Packaging Forum*.

50. Walker, M. (2017). *Why We Sleep: Unlocking the Power of Sleep and Dreams*. Scribner.

51. Chrousos, G. P. (2009). Stress and disorders of the stress system. *Nature Reviews Endocrinology*, 5(7), 374–381.

52. Chang, A. M., et al. (2015). Evening use of light-emitting eReaders negatively affects sleep, circadian timing, and next-morning alertness. *PNAS*, 112(4), 1232–1237.

53. Center for Environmental Health. (2021). Small swaps for big impact: Plastic detox guide.

54. Clear, J. (2018). *Atomic Habits*. Avery Publishing.

55. Earth Day Network. (2021). Building local environmental communities.

56. UNEP. (2019). *Celebrate Progress: Personal Sustainability Milestones*.

57. Bandura, A. (1986). *Social Foundations of Thought and Action: A Social Cognitive Theory*. Prentice-Hall.

58. Thompson, R. C., et al. (2009). Plastics, the environment and human health: Current consensus and future trends. *Philosophical Transactions of the Royal Society B*, 364(1526), 2153–2166.

59. Stephens, B., et al. (2013). Testing air cleaner effectiveness in reducing PM indoors. *Indoor Air*, 23(5), 379–386.

60. Offermann, F. J. (2009). Ventilation and indoor air quality in new homes. *California Air Resources Board Report*.

61. Mason, S. A., et al. (2018). Synthetic polymer contamination in bottled water. *Frontiers in Chemistry*, 6, 407.

62. Jiao, W., et al. (2016). Design and evaluation of a personal environmental monitor. *Sensors*, 16(12), 2042.

63. Prata, J. C., et al. (2020). Solutions and integrated strategies for microplastic pollution control. *Science of The Total Environment*, 736, 139680.

64. Napper, I. E., & Thompson, R. C. (2016). Release of synthetic microplastic fibers from domestic washing machines. *Marine Pollution Bulletin*, 112(1–2), 39–45.

65. PlanetCare. (2021). Product documentation and performance data.

66. De Falco, F., et al. (2019). Evaluation of microplastic release from textiles using washing simulation tests: Influence of textile characteristics. *Environmental Pollution*, 249, 534–539.

67. Morandim-Giannetti, A. A., et al. (2012). Biodegradable textile materials based on natural fibers and mycelium. *Journal of Cleaner Production*, 112(Part 4), 2823–2830.

68. Plastic Pollution Coalition. (2021). *How families can reduce plastic exposure*.

69. UNEP. (2019). *Reducing household plastic pollution: Air quality and home health*.

70. Environmental Working Group. (2020). *Personal care products and plastic contamination*.

71. Napper, I. E., & Thompson, R. C. (2016). Release of synthetic microplastic fibers from domestic washing machines. *Marine Pollution Bulletin*, 112(1–2), 39–45.

72. Nguyen Family Case Study. (2022). Community Sustainability Workshop, Ventura County, CA.

73. Ragusa, A., et al. (2021). Plasticenta: First evidence of microplastics in human placenta. *Environment International*, 146, 106274.

74. Li, D., et al. (2020). Microplastic release from infant feeding bottles. *Nature Food*, 1(11), 746–754.

75. Schwabl, P., et al. (2019). Detection of various microplastics in human stool: A pilot study. *Annals of Internal Medicine*, 171(7), 453–457.

76. Wright, S. L., & Kelly, F. J. (2017). Plastic and human health: A micro issue? *Environmental Science & Technology*, 51(12), 6634–6647.

77. Greenpeace. (2020). *Checklist for reducing plastic exposure at home*.

78. Jambeck, J. R., et al. (2015). Plastic waste inputs from land into the ocean. *Science*, 347(6223), 768–771.

Chapter 10 References:

1. Rochman, C. M., et al. (2016). The ecological impacts of marine debris: Unraveling the demonstrated evidence from what is perceived. *Ecology*, 97(2), 302–312.

2. Bergmann, M., et al. (2019). White and wonderful? Microplastics prevail in snow from the Alps to the Arctic. *Science Advances*, 5(8), eaax1157.

3. Geyer, R., Jambeck, J. R., & Law, K. L. (2017). Production, use, and fate of all plastics ever made. *Science Advances*, 3(7), e1700782.

4. NOAA Marine Debris Program. (2020). *Citizen Science Cleanup Guidelines*.

5. Keep America Beautiful. (2021). *Beach and river cleanup initiatives: Toolkit for volunteers*.

6. The Ocean Cleanup. (2023). *The Interceptor Project Overview*.

7. WWF. (2020). *Managing plastic waste in freshwater systems*.

8. Lebreton, L., et al. (2018). Evidence that the Great Pacific Garbage Patch is rapidly accumulating plastic. *Scientific Reports*, 8, 4666.

9. Ocean Voyages Institute. (2022). *Pacific Ocean cleanup voyage report.*

10. The Ocean Cleanup. (2021). *System 002 and gyre operations.*

11. Tsang, Y. F., et al. (2021). Technologies for microplastic removal from aquatic environments: A review. *Journal of Hazardous Materials*, 413, 125200.

12. Machado, A. A. d. S., et al. (2018). Microplastics as an emerging threat to terrestrial ecosystems. *Global Change Biology*, 24(4), 1405–1416.

13. European Bioplastics. (2022). *Biodegradable mulch films in agriculture.*

14. Lehmann, J., & Joseph, S. (2015). *Biochar for environmental management: Science, technology, and implementation.*

15. Composting Council Research & Education Foundation. (2020). *Best practices for plastic-free compost.*

16. Turner, A. (2018). Black plastics: Linear and circular economies, hazardous additives and marine pollution. *Environment International*, 117, 308–318.

17. Kumar, A., et al. (2021). Phytoremediation of microplastics-contaminated soil: A possible mechanism. *Chemosphere*, 279, 130578.

18. UNEP. (2019). *Land restoration strategies for plastic-polluted sites.*

19. City of San Francisco. (2022). *Plastic zoning ordinance and pilot remediation zones.*

20. Brahney, J., et al. (2020). Plastic rain in protected areas of the United States. *Science*, 368(6496), 1257–1260.

21. Kumar, R., et al. (2021). Vegetative barriers for air pollution control in urban environments. *Environmental Pollution*, 270, 116267.

22. Sherrington, C., et al. (2016). Investigating options for reducing releases in the aquatic environment from microplastics. *European Commission Report.*

23. Maiti, R., & Agrawal, S. (2020). Role of green roofs in mitigating urban air pollution. *Journal of Cleaner Production*, 258, 120577.

24. Dris, R., et al. (2016). Synthetic fibers in atmospheric fallout: A source of microplastics in the environment? *Marine Pollution Bulletin*, 104(1-2), 290–293.

25. Greenpeace International. (2013). *Toxic threads: The big fashion stitch-up.*

26. Silva, A. B., et al. (2018). Microplastics in the environment: Challenges and solutions. *Marine Pollution Bulletin*, 133, 336–345.

27. Lares, M., et al. (2018). Intercomparison of microplastic pollution in wastewater and treated effluent from sewage treatment plants. *Environmental Science and Pollution Research*, 25(6), 6126–6136.

28. Murphy, F., Ewins, C., Carbonnier, F., & Quinn, B. (2016). Wastewater treatment works (WwTW) as a source of microplastics in the aquatic environment. *Environmental Science & Technology*, 50(11), 5800–5808.

29. Ziajahromi, S., Neale, P. A., Rintoul, L., & Leusch, F. D. (2017). Wastewater treatment plants as a pathway for microplastics: Development of a new approach to sample wastewater-based microplastics. *Water Research*, 112, 93–99.

30. Rajala, A. Z., et al. (2021). Electrocoagulation of microplastics in water. *Journal of Environmental Chemical Engineering*, 9(1), 104571.

31. Larrain-Valenzuela, J., et al. (2021). Biofiltration technologies for removing microplastics in water: A review. *Science of the Total Environment*, 789, 147956.

32. ECHA (European Chemicals Agency). (2019). *Microplastics: Measures to reduce emissions.*

33. Sun, J., Dai, X., Wang, Q., van Loosdrecht, M. C., & Ni, B. J. (2019). Microplastics in wastewater treatment plants: Detection, occurrence and removal. *Water Research*, 152, 21–37.

34. French Parliament. (2020). *Microfiber filters mandatory on washing machines starting 2025.*

35. Li, J., et al. (2020). Microplastics in greywater: A source, pathway and removal analysis. *Science of the Total Environment*, 742, 140512.

36. UNEP. (2021). *Single-use plastics: A roadmap for sustainability.*

37. Beat the Microbead Campaign. (2023). *Global overview of microbead bans.*

38. OECD. (2021). *Extended Producer Responsibility and the environment: Policy approaches.*

39. European Commission. (2022). *Horizon 2020 program for plastic pollution R&D.*

40. Ellen MacArthur Foundation. (2017). *The new plastics economy: Catalyzing action.*

41. Plastic Collective. (2023). *Plastic offsetting: How it works and who it helps.*

42. UNEP. (2021). *Global Assessment of Plastic Pollution and Monitoring Systems.*

43. National Geographic Society. (2020). *Citizen science and plastic pollution: Global initiatives and impact.*

44. European Commission. (2022). *Real-time data platforms for environmental transparency.*

45. Break Free From Plastic. (2022). *Brand Audit Report: Top plastic polluters globally.*

46. GAIA (Global Alliance for Incinerator Alternatives). (2021). *Plastic waste characterization toolkit.*

47. Ocean Conservancy. (2023). *Community workshops for ocean plastic solutions.*

48. Earth Day Network. (2022). *Clean-up campaign guidelines and participation metrics.*

49. Plastic Pollution Coalition. (2021). *K-12 curriculum on plastic pollution.*

50. World Wildlife Fund. (2020). *Corporate partnerships in river restoration projects.*

51. Schmidt, C., et al. (2019). Plastic-to-fuel initiatives: Promise or peril? *Environmental Science & Policy*, 101, 230–239.

52. Biermann, L., et al. (2020). Finding plastic patches in coastal waters using optical satellite data. *Scientific Reports*, 10(1), 1–10.

53. da Costa, J. P., et al. (2019). Microplastic detection using fluorescent dyes: A review. *Science of the Total Environment*, 669, 435–449.

54. NOAA. (2022). *Smart buoys and monitoring networks for marine debris tracking.*

55. Yoshida, S. et al. (2016). A bacterium that degrades and assimilates poly(ethylene terephthalate). *Science*, 351(6278), 1196–1199.

56. Khan, S. et al. (2017). Biodegradation of plastics by fungi: A review. *Environmental Reviews*, 25(4), 396–408.

57. Tanasupawat, S. et al. (2020). Bioreactors for plastic biodegradation: Potential and challenges. *Biochemical Engineering Journal*, 158, 107568.

58. Liu, P. et al. (2019). Graphene oxide-based materials for removal of microplastics from aqueous environments. *Environmental Pollution*, 252, 1254–1261.

59. Wang, J. et al. (2020). Magnetic nanocomposites for microplastic removal: A new frontier. *Journal of Hazardous Materials*, 384, 121261.

60. Li, B. et al. (2021). MOFs as effective adsorbents for plastic additives in water systems. *Advanced Materials*, 33(19), 2007383.

61. Norwegian Institute for Water Research. (2021). *Microplastic removal efficiencies in municipal wastewater treatment plants.*

62. National Environmental Management Authority of Kenya. (2019). *Plastic ban progress report.*

63. Indian Ministry of Road Transport and Highways. (2022). *Plastic Waste Utilization in Road Construction.*

64. World Resources Institute Indonesia. (2021). *Community-based river clean-up strategies.*

65. Geyer, R. et al. (2017). Production, use, and fate of all plastics ever made. *Science Advances*, 3(7), e1700782.

66. Rochman, C. M., & Hoellein, T. (2020). The cost of cleaning up microplastics. *Nature Sustainability*, 3, 1020–1025.

67. PlasticsEurope. (2022). *Types of Plastics and Their Applications.*

68. Ellen MacArthur Foundation. (2021). *Closing the loop: Policy responses to plastic pollution.*

69. TerraChoice. (2010). *The sins of greenwashing: Home and family edition.*

Chapter 11 References:

1. Ellen MacArthur Foundation. (2016). *The New Plastics Economy: Rethinking the Future of Plastics.*

2. World Economic Forum. (2022). *Plastics and the Circular Economy.*

3. United Nations Environment Programme. (2023). *Towards a Global Treaty to End Plastic Pollution.*

4. United Nations Environment Programme. (2000). *The Montreal Protocol: Achievements and Challenges.*

5. Basel Convention Secretariat. (2019). *Amendments to Annexes II, VIII and IX related to plastic waste.*

6. Center for International Environmental Law. (2022). *Legal Pathways for a Binding Global Plastics Treaty.*

7. Earthwatch Institute. (2021). *Global Plastic Bans Tracker.*

8. OECD. (2021). *Extended Producer Responsibility: Updated Guidance for Efficient Waste Management.*

9. Rochester, J. R., & Bolden, A. L. (2015). Bisphenol S and F: A Systematic Review and Comparison of the Hormonal Activity of Bisphenol A Substitutes. *Environmental Health Perspectives*, 123(7), 643–650.

10. Lithner, D., Larsson, Å., & Dave, G. (2011). Environmental and health hazard ranking and assessment of plastic polymers based on chemical composition. *Science of the Total Environment*, 409(18), 3309–3324.

11. Myers, P. et al. (2011). *A Plastic Pandemic: Hidden Additives in Packaging*. Environmental Health Perspectives.

12. Ellen MacArthur Foundation. (2019). *Completing the Picture: How the Circular Economy Tackles Climate Change*.

13. Break Free From Plastic. (2021). *Brand Audit Report: The Top Corporate Plastic Polluters*.

14. Greenpeace. (2020). *The Myth of Plastic Offsetting*.

15. TerraCycle. (2023). *Loop: Reusable Packaging Platform Overview*.

16. Changing Markets Foundation. (2022). *Under Wraps: The Failure of Voluntary Corporate Plastic Commitments*.

17. EMF and IDEO. (2016). *The Circular Design Guide*.

18. World Bank. (2022). *Global Plastic Prices and Supply Chain Trends*.

19. UNEP. (2021). *Bioplastics and Their Role in a Circular Economy*.

20. European Bioplastics. (2023). *Bioplastics Market Data and Definitions*.

21. FAO. (2020). *Land Use Competition Between Bio-based Plastics and Food Security*.

22. CIEL. (2021). *Plastic & Health: The Hidden Costs of a Plastic Planet*.

23. Haneef, M. et al. (2017). *Advanced Materials from Mycelium-Based Bio-Composites*. Science Advances.

24. European Commission. (2022). *Future Bioplastics: Innovation and Scalability Challenges*.

25. Ellen MacArthur Foundation. (2020). *Upstream Innovation: A Guide to Packaging Solutions*.

26. World Economic Forum. (2021). *Supply Chain Redesign for Circularity*.

27. UNEP. (2022). *Plastic Pollution and Behavioral Change*.

28. Earthwatch Europe. (2021). *Plastic Investigators and Youth-Led Engagement*.

29. Eco-Schools International. (2023). *Impact Report on School Sustainability Programs*.

30. Social Media for Sustainability. (2021). *How Influencers Are Shaping the Zero-Waste Movement*.

31. BBC News. (2020). *Starbucks and Amazon Respond to Consumer Plastic Pressure*.

32. Ballotpedia. (2022). *Environmental Ballot Measures in the U.S.*

33. Sierra Club. (2021). *From Protest to Policy: Civic Action and Environmental Change*.

34. UNEA. (2023). *Toward a Global Agreement on Plastic Pollution*.

35. UN Environment Programme. (2022). *Draft Framework for Global Plastics Treaty*.

36. World Resources Institute. (2021). *National Alignment with Global Environmental Goals*.

37. CIEL. (2021). *Plastic Justice: Equity in the Global Plastics Treaty*.

38. WWF. (2022). *Case Studies in Plastic Innovation from the Global South*.

39. UN Environment Programme. (2023). *Global Treaty on Plastics: Draft Ratification Timeline*.

40. Ellen MacArthur Foundation. (2022). *Material Economics in the Circular Economy*.

41. OECD. (2023). *Policy Framework for Microplastic Mitigation*.

42. World Bank. (2024). *Projections on Plastic Leakage and Circularity*.

43. Center for the Circular Economy. (2021). *Product Design and Conscious Consumption*.

44. Journal of Environmental Ethics. (2020). *Reevaluating Convenience in Sustainable Cultures*.

45. McDonough, W., & Braungart, M. (2002). *Cradle to Cradle: Remaking the Way We Make Things*.

46. National Geographic. (2022). *How Public Awareness Is Fueling Plastic Reform*.

47. Plastic Pollution Coalition. (2021). *Imagining a World Without Plastic Waste*.

48. UNEP. (2023). *Call to Action: Global Stewardship for the Post-Plastic Age*.

Made in United States
Orlando, FL
05 November 2026

90231492R00167